Advances in Information Security

Volume 75

Series editor
Sushil Jajodia, George Mason University, Fairfax, VA, USA

More information about this series at http://www.springer.com/series/5576

Craig Rieger • Indrajit Ray • Quanyan Zhu
Michael A. Haney

Editors

Industrial Control Systems Security and Resiliency

Practice and Theory

 Springer

Editors
Craig Rieger
Critical Infrastructure Security
and Resilience
Idaho National Laboratory
Idaho Falls, ID, USA

Quanyan Zhu
Department of Electrical and Computer
Engineering
Tandon School of Engineering
New York University
Brooklyn, NY, USA

Indrajit Ray
Department of Computer Science
Colorado State University
Fort Collins, CO, USA

Michael A. Haney
Department of Computer Science
University of Idaho
Idaho Falls, ID, USA

ISSN 1568-2633
Advances in Information Security
ISBN 978-3-030-18213-7   ISBN 978-3-030-18214-4   (eBook)
https://doi.org/10.1007/978-3-030-18214-4

This Springer imprint is published by the registered company Springer Nature Switzerland AG
The registered company address is: Gewerbestrasse 11, 6330 Cham, Switzerland

Preface

While cybersecurity has been a consideration of information technologies (IT) for years, only since the last decade has an increase in concern for the security and resulting safety of our industrial control systems (ICS) been observed. Through standards and governmental agency guidance, the resources have been provided to better enable the asset owners to orchestrate better security architectures with current security technologies. Security vendors have advanced their product offerings to improve defenses against the evolving threat, and within the ICS community, ICS vendors have taken an active role to provide resources to the end users that enable consistent application and maintenance of cybersecurity. However, the threats that are specifically targeting ICS and the critical infrastructures we depend on are becoming more evident, as recognized by the HAVEX malware and others since then. Even with a consistent, risk-based application of security, an international challenge exists to evolve and transform the system architectures and technologies to be more resilient to cyber-threats.

As the desire to automate and achieve the efficiencies of labor and operation has grown, so has the investment in control systems that allow for integrating different operations, facilities, utilities, and infrastructures. Although significant strides have been made in making ICS secure, increasing the connectivity of systems with commodity IT devices and significant human interaction of ICS systems during its operation regularly introduces newer threats to these systems resulting in ICS security defenses always playing catch-up. To address this threat in the near-term solutions, the layers of protection that include those that are physically oriented, such as mechanically interlocking devices that have no cyber-connectivity, can reduce the risk associated with compromise of critical systems. However, as control systems evolve toward greater autonomy, reducing/changing the role of the human, the need to consider resilience becomes more profound. Autonomous systems can react quickly to anomalous conditions, ensuring we have power even if a transformer fails. However, it can also cause a quick escalation to a cascading fault if the autonomy has been corrupted by cyber-attack or unrecognized failure.

The next generation of control systems should have a better understanding of threat versus quality-of-service trade-offs. Reasoned by such trade-offs, the next-generation control systems should be designed to be resilient by nature. Such resilient ICS design requires one to be proactive in understanding and reasoning about the relationships and dependencies between the various ICS components, evolving threats to them, and the effects of these threats on the mission goals of the ICS system. As such, the ability to not only detect but correlate the impact on the ability to achieve minimum normalcy is a necessary attribute. Enabling the human in the loop will be necessary throughout, ensuring their ability to adapt to anomalous conditions that the control system cannot. Threat-resilient architectures will provide a holistic feedback and data-driven security solution that integrates a real-time cyber-physical risk assessment, proactive and adaptive defense mechanism, and decentralized reconfigurable resilient control design. The risk assessment evaluates the real-time risks at the cyber and physical components of the system that can provide reliable information for defense and control systems to respond.

Autonomous proactive defense mechanisms, such as deception and moving target defenses, are pivotal to strategically adapt to adversarial behaviors, create information asymmetry to deter the attacks, reduce attacker's advantage, and mitigate the losses. The resilient control design is the last mile protection for the industrial control systems. A resilient controller can reconfigure the physical layer control laws that can steer the control system away from the damages through quick detection, failure localization, and fast response in a distributed fashion. The integrated design of risk measure and learning, autonomous defense, and resilient controls plays an important role in improving the resiliency of the system holistically. Resilience measures provide quantitative metrics to guide the design process to achieve desirable system-level performance. Multidimensional metrics, such as response time and loss of performance, at both cyber and physical layers of the ICS are important indicators and need to be part of the design goals of the next-generation architectures.

In this edited volume, we hope to provide different perspectives for achieving near- and long-term resilience, including technologies of the future. Therefore, what follows is a synopsis of the current challenges that will need to be addressed in future control systems designs. Current automation environments are the result of organic interconnection of control systems and the inability to recognize and prevent resulting, unrecognized faults. Addressing near-term resilience in this context requires an understanding of the consequence and efficient use of resources to address. In moving toward inherent resilience, adaptive and agile distributed frameworks for recognizing and responding to threat are necessary. Benign human error as the result of data overload and lack of information is an ongoing issue, and for the malicious human, current perimeter protections are insufficient and not designed to adapt rapidly to attacks in order to prevent compromise. The development of autonomous defenses that use the attackers' humanness against them is an imperative. Finally, current control systems have multiple performance goals, but without the necessary identification and prioritization can lead to undesirable response from both the human operation and the automation design. Enabling the success of the operator requires integration of visualizations, such that the various roles of cyber-defender or process operator can maintain the same context, for the former an understanding of what is important in the process and the latter how cyber-assets are affecting the physical operation.

Idaho Falls, ID, USA   Craig Rieger

Contents

Part I
Current and New Practice

Current Standards for Cyber-Hygiene in Industrial Control System Environments

Ken Modeste

Abstract Industrial control systems (ICS) have historically been closed systems reliant on serial connectivity that was exclusive to these networks. The potential for cybersecurity incidents associated with these closed systems required physical access to the facilities and hence was considered low risk in most circumstances.

Introduction

Industrial control systems (ICS) have historically been closed systems reliant on serial connectivity that was exclusive to these networks. The potential for cybersecurity incidents associated with these closed systems required physical access to the facilities and hence was considered low risk in most circumstances.

However, as technology has rapidly started adapting to a newer world of connectivity, Internet of things (IoT), and cloud systems, the potential for ICS connectivity to information technology (IT) systems, and general trends to IoT, these systems have been migrating to open systems that are connected via Ethernet or wireless to the rest of the commercial use networks in facilities. As such, these open networks are now being connected to the Internet for a multitude of innovative and new capabilities, driving some areas such as:

(a) Remote maintenance and diagnostics of facility equipment
(b) Data collection and analytics
(c) Cloud service capabilities
(d) Smart systems with aggregation of sensor data for business analytics

Vendors, system installers, operators, and facility owners now have newer capabilities that promote economic value and technology upgrades that align with twenty-first-century opportunities and competitiveness. The traditional and managed concepts of safety that covered hazards like fire, electric shock, or person harm now

K. Modeste (✉)
Underwriters Laboratories, Northbrook, IL, USA
e-mail: Ken.Modeste@ul.com

© Springer Nature Switzerland AG 2019
C. Rieger et al. (eds.), *Industrial Control Systems Security and Resiliency*, Advances in Information Security 75, https://doi.org/10.1007/978-3-030-18214-4_1

have additional risks with this new connectivity to other commercial and enterprise systems and the Internet. These new risks to safety can now be classified with disruption of businesses; additional risks to new safety concerns like privacy, exfiltration of data, remote control, and modification of equipment outside of their intended use; and ultimately use of ICS equipment and systems for unplanned nefarious purposes.

Incorporating new cyber-technologies, methods, and processes in the design, development, installation, support, and use of ICS equipment requires standardization to support the industry in applying best practices that are economically feasible, relevant, and capable of assessing and managing these risks. Understanding the relevant standards and specifications available that can be applied to the ICS industry can support all stakeholders in continuing to apply new and innovative technologies that address connectivity and IoT opportunities while effectively managing the associated risk.

Ways to Address Cyber-Hygiene

Consider cyber-hygiene similarly to our own personal bodies and health hygiene practices. As personal hygiene revolves around activities that individuals incorporate into their regular practices, cyber-hygiene does the same. What are the best practices that organizations can deploy to continue to maintain the organization's cyber-well-being or improve upon it?

These best practices can involve the following common solution areas:

1. *Design specifications and standards*

    These standards and specifications help manufacturers by providing guidance in how to implement cybersecurity controls in products, components, and systems in aligned industries. These design standards may also apply to specific technologies that implement good cyber-capabilities (i.e., cryptography, software updates, etc.).

2. *Test and performance standards*

    These standards provide capabilities to evaluate and assess cybersecurity capabilities in products, components, and systems. Typically, they are used by trusted third parties to evaluate, assess, or audit cybersecurity practices, or can be used to assess a design standard.

3. *Product development team processes*

    Frameworks that define the process used to build products from their inception to their eventual decommissioning. These processes incorporate cybersecurity features from the beginning to ensure a vendor's cybersecurity objectives are built into the development process.

4. *Organization and process standards*

    Audit criteria for assessing an organization's overall cybersecurity practices. Vendors, system installers, and building owners have standard operating procedures that cover their business practices.

5. *Personnel training*

These standards provide the criteria for a person to be evaluated for their qualifications to support cybersecurity capabilities in the ICS space.

6. *General*

These standards and specifications typically will define technologies and provide general system descriptions and overall technical guidance on how particular technologies operate.

Standards

North American Electric Reliability Corporation Critical Infrastructure Protection (NERC CIP)

One of the more well-known standards are the NERC CIP[1] series of standards for physical security and cybersecurity. These standards provide minimum security requirements for bulk power generation in the USA, Canada, and parts of Mexico. These standards were adopted in 2006 and are defined in Table 1.

The NERC standards provide for a comprehensive cybersecurity framework. These standards are considered typical in an audit process to confirm that the policies and procedures in place can provide a minimum level of security for BES. There are also some associated security best practices that can be located at https://www.eisac.com/resources/documents.

ISA/IEC 62443

ISA99 is the name of the Industrial Automation and Control System (IACS) Security Committee of the ISA.[2] This committee developed the series of ISA 62443 standards and technical reports. The intended audience for this series of standards and technical reports are asset owners, system operators and integrators, and ICS manufacturers. It is intended to provide guidance that an asset owner can use as procurement criteria for its supply chain and system operators to follow. These standards are now being reviewed and published as IEC 62443 in conjunction with IEC Technical Committee 65 Working Group 10 (IEC TC68/WG 10) as international standards. They fall into four categories, as defined in Table 2 for IEC publications.

These standards are well known in the factory automation space and are seeing some traction in the oil and gas market. They are designed specifically to ensure an asset owner can define cybersecurity objectives for its automation facility with a

---

[1]https://www.nerc.com/pa/Stand/Pages/CIPStandards.aspx

[2]ISA, International Society of Automation (http://www.isa.org/)

Table 1 NERC CIP standards

Standard	Type	Description
CIP-002	Design specifications and standards	*Bulk Energy System (BES) Cyber-System Categorization* Provides criteria for the inventory of device and software assets of a BES that can adversely impact the reliability of the BES via a regular risk assessment methodology
CIP-003	Organization and process standards	Policies for security management controls to prevent compromise of the BES
CIP-004	Personnel training	Security awareness and training for personnel operating and managing BES
CIP-005	Design specifications and standards	Electronic security perimeter controls for a BES
CIP-006	Design specifications and standards	Physical security controls for a BES
CIP-007	Design specifications and standards	System security management of BES, which defines the security controls of the system, how to assess those controls, and continuous vulnerability management
CIP-008	Organization and process standards	Incident response and planning policies for a BES
CIP-009	Organization and process standards	Recovery plans for a BES in the event of a shutdown, failure of controls, or a cyber-event
CIP-010	Organization and process standards	Configuration change management policies for a BES
CIP-011	Organization and process standards	Policies and procedures for information protection of a BES

defined cybersecurity maturity level; manufacturers can then design to those requirements to meet the security level prescribed at the maturity level. System installers, integrators, and service providers can then be trained on the established objectives of the asset owner and the implementation knowledge of the components to meet those objectives.

Underwriters Laboratories (UL) 2900

These standards were developed to provide testing criteria for product components and systems. The UL 2900 series focuses on the best cybersecurity practices that are used in assessing devices and components when addressing the software and firmware. They fall into three categories as defined in Table 3 for the ANSI/CAN/UL publications.

The UL 2900 series of standards are designed to provide testing criteria to evaluate and assess manufacturer's devices, components, and ICS. Its targeted audience are asset owners to use as procurement requirements for ICS manufacturers to meet for third-party testing and certification and for ICS manufacturers to use on their supply chain.

Table 2 IEC publications

Standard	Type	Description
Part 1: series of the standards covers general terms, glossary items, and ICS life cycle and use cases		
62443-1-1	General	Defines terminology, concepts, and models for IACS typically used in factory automation This section also defines seven functional requirements in securing an IACS, which are: (a) Identification and authentication control (b) Use control (c) System integrity (d) Data confidentiality (e) Restricted data flow (f) Timely response to events (g) Resource availability
Part 2: series of the standards covers policies and procedures for an asset owner or system operator		
62443-2-1	Organization and process standards	*Industrial communication networks: network and system security* *Part 2-1: establishing an industrial automation and control system security program* This standard provides guidance for application of a cybersecurity management system for IACS systems and is based on ISO/IEC 17799 information technology – security techniques – code of practice for information security management and ISO/IEC 27001 standards information technology – security techniques – information security management systems: requirements that describe a cybersecurity management system for business/information technology systems
62443-2-3	Organization and process standards	Technical report for patch management of an IACS system
62443-2-4	Organization and process standards	Security program requirements for IACS service providers. This standard introduces the four maturity levels of an organization. These security levels are based on the maturity levels found in Capability Maturity Model Integration (CMMI)[a] for services called CMMI-SVC. The levels are used throughout the series of the standard as they define for an asset owner where an expectation of capabilities and risk management exists
Part 3: series of the standards covers policies and procedures for an system operator, installer, and integrator		
62443-3-1	General	*Industrial communication networks – network and system security – Part 3-1: security technologies for industrial automation and control systems* This defines the typical technologies that would exist to promote security in an IACS

(continued)

Table 2 (continued)

Standard	Type	Description
62443-3-3	Design specifications and standards	*Industrial communication networks – network and system security – Part 3-3: system security requirements and security levels* Taking the seven functional requirements in 62443-1-1, this standard defines four security level requirements for each of the functional requirements from one to four with increasing levels of security based on the risk of exposure to the IACS based on an attackers capabilities and means
Part 4: series of the standards covers policies and procedures for a manufacturer of IACS components		
62443-4-1	Organization and process standards	Secure product development life cycle requirements These requirements provide criteria for a manufacturer of IACS components to follow when designing and building the IACS component. They are aligned with industry best practices around secure development life cycles (SDL)
62443-4-2	Design specifications and standards	Technical security requirements for IACS components To specify security capabilities that enable a component to be integrated into a system environment at a given security level An ICS component shall be designed for relevant requirements of this standard per the security level where the ICS component is intended to be installed

[a]https://cmmiinstitute.com/

Table 3 UL standards

Standard	Type	Description
Part 1: series of the standards covers the general requirements to assess any product, device, component, or system when addressing the software and firmware risks		
UL 2900-1	Test and performance standards	*Software cybersecurity for network-connectable products* *Part 1: general requirements* These requirements provide testing criteria for any device that contains software or firmware
Part 2: series of the standards covers industry-specific requirements		
UL 2900-2-2	Test and performance standards	*Software cybersecurity for network-connectable products* *Part 2-2: particular requirements forICS* These requirements provide testing criteria for any ICS, devices, or components that contain software or firmware

Table 4 NIST standards

Standard	Type	Description
SP 800-53	Organization and process standards	*Security and privacy controls for federal information systems and organizations* They provide guidelines for selecting and specifying security controls for organizations and information systems. The IEC 62443 and UL 2900 security controls typically follow this popular guidance document
SP 800-53A	Test and performance standards	*Guide for assessing security controls in information systems* They provide assessment criteria for SP 800-53
SP 800-82	General	*Guide to ICS security* Provides guidance for securing ICS, supervisory control and data acquisition (SCADA) systems, distributed control systems (DCS), and other systems performing control functions
SP 800-94	General	*Guide to intrusion detection and prevention systems* This can be used by a system installer or operator to provide guidance on how to configure and set up intrusion detection and prevention systems
SP 800-87	General	*Establishing wireless robust security networks* Provides good guidance on setup and configuration of wireless networks following the IEEE 802.11i-based wireless local area networks (LANs)
NIST cybersecurity framework	Organization and process standards	*Framework for improving critical infrastructure cybersecurity* Provides capabilities for organizational assessments of critical infrastructure assets. The model is based on five major tenets: identify, protect, detect, respond, and recover
NIST IR 7176	Test and performance standards	*Protection profile for ICS*

National Institute for Standards and Publications (NIST) Special Publications

The National Institute for Standards and Publications (NIST)[3] of the US government produces many specifications to provide guidance and best practices for use in critical infrastructure. These are referenced fairly prolifically throughout the industry and begin from an overall description of ICS and how to implement security all toward the specifics needed to define robust cybersecurity practices and are defined in Table 4.

---

[3]https://csrc.nist.gov/publications/sp

Department of Homeland Security (DHS) and Department of Energy (DOE) Publications

The US Department of Energy (DOE) produced a capability maturity model through the Cybersecurity Capability Maturity Model (C2M2) program.[4] C2M2 focused on the implementation and management of cybersecurity practices associated with the operation and use of information technology and operational technology assets and the environments in which they operate. The goal of these maturity models was to provide clarity in general and for certain sectors like electricity and oil and gas for asset owners and system operators to determine a baseline of where their current cybersecurity practices are and to develop goals for cybersecurity objectives in the future.

The US Department of Homeland Security (DHS) and its Industrial Control System Cyber Emergency Response Team (ICS-CERT)[5] continually work to address challenges and risks within ICS regarding cybersecurity. The Common Criteria for Information Technology Security Evaluation is a program mutually recognized by 28 countries worldwide that uses the technical standard ISO 15408 *information technology – security techniques – evaluation criteria for IT security* as a foundation for developing security requirements for a particular system or device. The evaluation criteria are developed in associated protection profiles. NIST produced an NIST Interagency Report (IR) called NIST IR 7176, which provides a protection profile that document security requirements associated with ICS.

DHS also produces several documents and specifications that educate the industry on best practices, ongoing risk mitigation techniques, and general good hygiene for the industry, which are defined in Table 5.

Smart Grid Publications

There are several standards that focus on helping manufacturers design equipment specific in the smart grid space. These standards are typically focused on specific types of equipment and their use, or communication protocols in the smart grid, and how to deliver security requirements into the protocol. They are defined in Table 6.

---

[4]https://www.energy.gov/oe/cybersecurity-critical-energy-infrastructure/cybersecurity-capability-maturity-model-c2m2-program

[5]https://ics-cert.us-cert.gov/Standards-and-References

Table 5 DHS documents

Standard	Type	Description
ES-C2M2	Organization and process standards	*Electricity Subsector Cybersecurity Capability Maturity Model* This specification covers a common set of industry acceptable best cybersecurity practices that cover the electricity subsector
ONG-C2M2	Organization and process standards	*Oil and Natural Gas Subsector Cybersecurity Capability Maturity Model* This specification covers a common set of industry acceptable best cybersecurity practices that cover the oil and gas subsector
Control system catalog	Organization and process standards	*Catalog of Control Systems Security: Recommendations for Standards Developers* It specifies a catalog of security controls applicable to ICS from different standards, specifications, and other industry publications
Control system cybersecurity	Organization and process standards	*Recommended Practice: Improving Industrial Control System Cybersecurity with Defense-in-Depth Strategies* Provides a good overview of deploying defense in depth for an ICS
Procurement language	Organization and process standards	*Cybersecurity Procurement Language for Control Systems* Provides security principles for ICS when considering designing and acquiring ICS
CNSSI-1253R2	Organization and process standards	*Security Categorization and Control Selection for National Security Systems* This document uses NIST SP 800-53 and establishes the processes for categorizing facilities and the information they process and for appropriately selecting security controls from NIST SP 800-53
CNSSI-1253	Organization and process standards	*Security control overlays for ICS* Specifications of security controls and supporting guidance used to complement the security control baselines and parameter values in the supplemental guidance in NIST SP 800-53

French Network and Information Security Agency (ANSSI)

The French government, through its security agency, ANSSI,[6] has been producing standards and specifications for critical infrastructure to subject all new critical ICSs to an approval process, thus ensuring that their cybersecurity level is acceptable given the current threat status and its potential developments. Some of those produced recently and are becoming commonplace in new deployments in France are shown in Table 7.

---

[6]https://www.ssi.gouv.fr/publications/

Table 6 Smart grid publications

Standard	Type	Description
IEEE 1686	Design specifications and standards	*Substation Intelligent Electronic Devices (IEDs) Cybersecurity Capabilities* Covers applying security controls to IEDs regarding the access, operation, configuration, firmware revision, and data retrieval
IEEE C37.240	Design specifications and standards	*Cybersecurity Requirements for Substation Automation, Protection, and Control Systems* Covers security controls implemented at the substation that factors in risk levels associated with the business practice and the cost associated with the technical control
NISTIR 7628	Organization and process standards	*Guidelines for Smart Grid Cybersecurity* Provides best practices for an asset owner deploying smart grid technology to consider security implications

Table 7 ANSSI standards

Standard	Type	Description
Cybersecurity for ICS	Organization and process standards	*Classification method and key measures* Provides a mechanism to classify ICS based on acceptable risk and how to measure the classes defined
Cybersecurity for ICS	Organization and process standards	*Detailed measures* Provides technical and organizational criteria needed for cybersecurity for new ICS systems that fall under industry 4.0[a]

[a]Industry 4.0 is a European focus of industrial Internet of things where ICS systems are integrated to external systems via the Internet

Bundesamt für Sicherheit in der Informationstechnik (BSI)

The German Federal Office for Information Security (BSI) has been developing standards and best practices around industry 4.0 and cybersecurity principles necessary for the German economy. The German government recently launched a cybersecurity implementation plan for critical infrastructure called KRITIS,[7] primarily intended to focus on securing the country's networked information infrastructure while making it still productive and economically competitive. KRITIS is Germany's contribution to the European Program for Critical Infrastructure Protection (EPCIP). Some of these specifications can be seen in Table 8.

The industrial Internet of Things Consortium[8] has developed several technical documents to help instruct industry on the risks and challenges in having IoT and ICS. They have published an Industrial Internet of Things Security Framework, which provides some of the general understanding of how the industrial Internet

---

[7]https://www.kritis.bund.de/SubSites/Kritis/EN/strategy/strategy_node.html

[8]http://www.iiconsortium.org

Table 8 BSI specifications

Standard	Type	Description
ICS Security Compendium	General	*ICS Security Compendium* This is a great reference document that outlines the security in ICS procedures and the relevant standards globally that can support
CIP Implementation Plan	General	*CIP Implementation Plan of the National Plan for Information Infrastructure Protection* Provides a national plan for securing the national information technology infrastructure based on prevention, preparedness, and sustainability
Baseline protection concept	Organization and process standards	*Protection of Critical Infrastructures: Baseline Protection Concept* Provides facilities based in Germany with guidelines for the internal cybersecurity of the facility

Table 9 Personnel training specifications

Standard	Type	Description
CompTIA	Personnel training	CompTIA[a] has several certifications with criteria for qualification around general cybersecurity, cloud systems, and security testing
EC-Council	Personnel training	EC-Council[b] has several training and certification programs with popularity around the ethical hacker courses
GIAC	Personnel training	GIAC[c] has several standard technology certification programs and specific criteria for ICS personnel
ISACA	Personnel training	ISACA[d] focuses on training personnel for specific cybersecurity roles within an organization
(ISC)2	Personnel training	(ISC)2[e] qualifies different roles in cybersecurity and the required credentials

[a]https://certification.comptia.org/certifications
[b]https://www.eccouncil.org/programs/
[c]https://www.giac.org/
[d]http://www.isaca.org/Certification/Pages/default.aspx
[e]https://www.isc2.org/Certifications

would technically be deployed and some of the main elements needed to ensure the security of such a deployment.

Personnel Training

Ensuring the personnel that design, build, manufacture, install, service, and operate critical infrastructure systems supports the general cyber-hygiene of an overall system. Qualified personnel who have capabilities to support the cyber-objectives of an installation drive overall competency. Some of those certified specifications are shown in Table 9.

Summary

This chapter provided a list of cybersecurity standards and specifications, which can help in developing a good way to determine cyber-hygiene in critical infrastructure systems. However, one singular standard or specification cannot provide a truly holistic view of the cyber-capabilities of a facility's implementation of systems and services. A combination of several "types" of standards would provide the best avenue to ensure that an organization is using the best capabilities readily available.

One of the first steps to help an asset owner determine this is to understand the nature of some of the cybersecurity and critical infrastructure risks. Several "general" standards can provide great insight for someone who is attempting to understand the landscape of a system. NIST SP 800-82 and the Industrial Internet Security Framework are both good places to start to get a good declaration on control systems and what is typically done to secure them.

The asset owner would then need to assess the current state of his/her system by using some of the identified "organization and process standards." These standards, like the NIST Cybersecurity Framework, DOE's Capability Maturity Model, or ANSSI's cybersecurity for ICS, can provide an overall assessment of the current state of his/her system. Included in that is the need to examine the relevant staff charged with maintaining those systems and ensure they have the relevant credentials to execute on cybersecurity-related activities. Using some of the "personnel training" standards to assist candidates of the asset owner's technical staff to increase their knowledge can help as well. Cybersecurity professionals can either learn on the job or be trained beforehand. Understanding the current state and capabilities of one's current staff will provide an asset owner with a good understanding of where his/her organization currently is.

The next step is to use some of the identified "organization and process standards" to build a scalable plan to help identify a target or desirable state of the facility's completed cybersecurity capabilities. A capability maturity model can help set up target capabilities and create a plan to get there. Using some of the "test and performance standards," in combination with the "organization and process standards," can provide asset owners with a way to measure how good the current facility is. NIST SP 800-53A, combined with IEC 62443-2-4 (which takes much of its input from NIST SP 800-53A), can evaluate the current state. This is what the NERC CIP standards in the bulk energy sector focus on by providing the criteria needed to perform an assessment of what an organization has built into its infrastructure to meet cybersecurity requirements. Using procurement guidance to begin building procurement requirements for the supply chain of the facility would be another great step by informing system operators, installers, and maintenance teams of control systems, integrated technology systems, etc., of what is expected of them. The qualifications in installing and servicing equipment to make sure they meet a manufacturer's stated specifications are crucial in meeting cybersecurity needs.

Procurement language can also be driven into the entire supply chain of the infrastructure. "Design specifications and standards" and "test and performance standards" can then be used to document what criteria are needed for equipment and services and how those systems will be assessed. The design standards will provide technical criteria that must be met for a device, component, or system to be acquired, and the test standards can provide compliance criteria to evaluate and assess those capabilities. In this regard, the 62443-3-3, 62443-4-2, and DHS cybersecurity control documents can provide information to the supply chain of the technical security controls that are needed. UL 2900 can be used to evaluate and assess the supply chain's devices, components, and systems, so a procurer can expect a trusted third party to perform assessments and provide a certified and qualified supply chain.

Manufacturers of the supply chain can then apply these "design specifications and standards" and "test and performance standards" to build the products to be used in the installation. Manufacturers in the supply chain can also apply "product development team processes" standards to ensure security is considered when building those products. These standards would focus on driving some of the best practices developed by leading organizations in delivering quality products and systems designed with cybersecurity risks in mind for the impacted product in certain implementations and factor in mitigation and control capabilities to minimize those risks. Manufacturers can even apply the same "organization and process standards" to their organizations as well to robustly build a team that can address security risks both inside the organization and for the processes used to build products for the industry. Ultimately, the manufacturer can apply the "personnel training" standards to qualify their technical resources in building their products, by pushing them through their own supply chains.

As has been demonstrated, asset owners can use an amalgam of these standards and specifications to provide robust capabilities for their systems. Most of these standards align with common best practices for systems in critical infrastructure globally, and are recognized by industry and cybersecurity professionals. Once maturity levels are defined, and plans are made to ascertain a certain level, the right standards, specifications, and guidance documents will align with an asset owner's cybersecurity plans.

Consequence-Based Resilient Architectures

Curtis St. Michel and Sarah Freeman

Abstract As described in Lee et al., cyber-attackers conducted a coordinated, multifaceted operation against three distribution companies on 23 December 2015, resulting in a customer outage of nearly 4 hours. The significance in this event does not originate from the infiltration of the electric sector; on the contrary, Gorman, Toppa, Perlroth, Dearden, and Borger indicate they have been compromised before and will continue to be compromised in the future. Nor was this event significant because it harkened the arrival of some previously unknown, sophisticated industrial control system (ICS) malware, as Karnouskos, Fidler and Matrosov et al. argued was the case with Stuxnet. Rather, the significance of the December 2015 event stems from the means by which the attackers interfaced with and, ultimately, used the energy system design to their advantage.

The Challenges of Security by Design

As described in Lee et al. [1], cyber-attackers conducted a coordinated, multifaceted operation against three distribution companies on 23 December 2015, resulting in a customer outage of nearly 4 hours. The significance in this event does not originate from the infiltration of the electric sector; on the contrary, Gorman [2], Toppa [3], Perlroth [4], Dearden [5], and Borger [6] indicate they have been compromised before and will continue to be compromised in the future. Nor was this event significant because it harkened the arrival of some previously unknown, sophisticated industrial control system (ICS) malware, as Karnouskos [7], Fidler [8] and Matrosov et al. [9] argued was the case with Stuxnet. Rather, the significance of the December 2015 event stems from the means by which the attackers interfaced with and, ultimately, used the energy system design to their advantage.

Engineering controls are the result of countless hours of analysis, during which design engineers validate the safety, reliability, and functionality of a designed

C. St. Michel (✉) · S. Freeman (✉)
Control Systems Cybersecurity Analyst, Idaho National Laboratory, Idaho Falls, ID, USA
e-mail: Curtis.StMichel@inl.gov; Sarah.Freeman@inl.gov

© Springer Nature Switzerland AG 2019   17
C. Rieger et al. (eds.), *Industrial Control Systems Security and Resiliency*, Advances
in Information Security 75, https://doi.org/10.1007/978-3-030-18214-4_2

system. One prevalent method for validation is failure mode and effects analysis (FMEA), a systematic approach for proactively identifying where and how a system might fail, as well as any potential resulting impact. FMEA and its variants, such as failure mode, effects, and criticality analysis (FMECA) and multi-attribute failure mode analysis (MAFMA) [10], are linked in their failure to properly consider cyber-events and their potential impact to reliability and, ultimately, the resiliency of a designed system.

As additional digital components have been introduced into traditionally analog systems, the risk associated with equipment failure shifts. This is due in part to a change in the device control themselves, as well as the possibility for additional malicious activity directed against this equipment. For example, cyber-attacks can be multiplied by employing attacks that both rely on the visibility digital sensors and data aggregators, as well as the manipulation of engineering control algorithms themselves [11]. Although these changes in technology can provide a wealth of data management opportunities and improved efficiency, this shift has also posed a challenge for individuals and organizations tasked with securing this equipment.

The shift toward an increased reliance on digital technology harkens the arrival of a new reality in which these systems and technology can be used for increasingly sophisticated cyber-attacks. Events against electric grids worldwide since 2015 highlight the distinct difference in targeted and untargeted cyber-attacks and the failure of perimeter cyber-defense to combat directed attacks. Today traditional cyber-hygiene and best practices, although important, are no longer sufficient to stop targeted cyber-attacks. At the same time, traditional FMEA and its variants must evolve to address both adversary capability and consumer demand for technology so that reliability, safety, and resiliency of these critical engineered systems continue.

The Vulnerability Mitigation Cycle

Vulnerability assessments are a requirement for North American Electric Reliability Corporation (NERC) Critical Infrastructure Protection (CIP) compliance and are intended to limit the possibility of a cyber-attack against the bulk electric system (BES); numerous guides have been written, and research has been conducted to optimize these activities, most notably by Sandia National Laboratories [12]; Ten et al. [13]; and Ralston et al. [14]. This technique is fundamentally limited to known vulnerabilities or the zero-day vulnerabilities that may be found by a cybersecurity researcher as part of that assessment, however. Additionally, organizations and vendors frequently also employ a vulnerability/mitigation strategy that involves the application of patches as new vulnerabilities become known. The fundamental result of this system is one in which the individual vulnerabilities that are identified and mitigated focus primarily on known adversary capabilities and exploits. Therefore, a proactive vulnerability management strategy becomes inherently reactive.

Unfortunately, given the speed at which new vulnerabilities are identified, organizations face an uphill battle in securing their operational technology (OT) space.

Vulnerabilities with some of the greatest potential for weaponization, zero days, are so named due to the fact that they are vulnerabilities in systems that were otherwise unknown, with no patching available at the time of their discovery. In 2013, the number of zero-day (0-day) vulnerabilities discovered doubled from the previous year to 23. Between 2014 and 2015, there was a 125% increase in the number of vulnerabilities to 52, leading Symantec [15] to theorize that zero days have been "professionalized," a critical tool for state-sponsored activity. Security is also complicated by the white phase of the zero-day life cycle, when a patch has been released but in many cases has not yet been applied. Dacier et al. [16] noted a five times increase in the malicious use of zero-day vulnerabilities, after they had been disclosed, highlighting the continued risk posed to organizations even later in the vulnerability life cycle. This finding is shared by Ablon and Bogart [17], a recent review that evaluated more than 200 zero days over 14 years (2002–2016), which found that the average "life expectancy" following discovery of a zero-day vulnerability averaged 6.9 years. In 25% of these cases, life expectancy for these vulnerabilities averaged more than 9.5 years. Within the ICS/Supervisory Control and Data Acquisition (SCADA) space, where patches occur far less frequently, it is possible that life expectancy is even higher.

In general, OT is patched far less frequently than its information technology (IT) cousin. Tom et al. [18] note that legacy systems are typically patched late, if patched at all, in part due to "...their service age, proprietary nature, perceived obsolescence, or simply because the patches are unavailable." The result is that vulnerabilities, zero days or not, can be used to exploit OT for several years and, given the rate at which new vulnerabilities emerge and the lack of infinite resources to devote to cybersecurity, complete mitigation through patching cannot be expected.

How, then, can organizations protect themselves from the inevitable stream of vulnerabilities? The best approach may be not to focus on the vulnerabilities, but to introduce resiliency into the technical designs themselves, through methods of consequence-based analysis.

Consequence-Driven Cyber-Informed Engineering

Introduced by St. Michel et al. [19], Consequence-driven Cyber-informed Engineering (CCE) is one method to address the organizational risk posed by increasingly sophisticated cyber-attacks. Rather than focus solely on the vulnerability mitigation cycle, CCE prioritizes cybersecurity response capabilities based on impact and, ultimately, the potential severity of a cyber-attack. In this way, CCE addresses the most significant threat to an organization's critical functions and services in a resource-constrained environment.

The motivation for the development of CCE stemmed in part from the development of increasingly sophisticated adversary capabilities and the corresponding challenges associated with the vulnerability mitigation cycle. CCE also originated from the increasingly prevalent (if not pessimistic) view that perfect (or event near perfect) cybersecurity protection is a mirage and something that cannot be realistically achieved [20]. If this view is to be adopted, then any organization is limited in its ability to develop suitable responses to the threat of cyber-attack. In many cases, the challenge of securing critical systems, processes, and procedures from a sophisticated, targeted, state-sponsored cyber-attack exceeds the capabilities of the organization.

The problem of security is compounded by the increasingly varied cyber-boundaries of an organization. An electric utility, for example, expects to exchange some amount of operational information with other utilities, especially those whose infrastructure they interface with or with whom they conduct electricity market transactions.

The cyber-boundary has also shifted through the adoption of emerging technology. In late March–early April 2018, three US pipeline companies experienced communications system disruptions after a third-party provider experienced a cyber-attack [21]. The affected system existed on the boundary of the organizations, and although it did assist with operational activity – by providing communications support to the pipeline customers and their purchases – it did not fundamentally inhibit the delivery of product. Operationally, transactions were able to continue, albeit at a slower pace. Still, the event highlights cybersecurity challenges. Traditional definitions and boundaries of the electronic perimeter have become obfuscated by technology intended to improve or streamline operational activity.

If perfect or near perfect cybersecurity is not a possibility, how can organizations respond to the potential risk of a cyber-attack? One idea gaining additional traction is the concept of cyber-insurance, a risk management approach in which the individual or organization provides an insurance premium to transfer the risk to an insurance company [20]. In the event of a cyber-attack, the cost of the event would be distributed among the collective pool of individuals and organizations purchasing insurance. The market for cyber-insurance continues to grow; Romanosky et al. [22] note that with less than $1 billion in premiums in 2012, estimates are as high as $20 billion by 2020. Still several barriers to an effective cyber-insurance market persist. For example, in spite of the growth, the cost associated with a cyber-event greatly outweighs the cyber-insurance market with estimated global costs of $445 billion a year [23].

Another issue for insurance companies is how to underwrite and define the risk they are willing to absorb and the cyber-incidents they are willing to cover. Related questions include how to quantify an organization's protection or exposure and what cybersecurity components, exactly, are the responsibility of a single organization? What degree of protection is a reasonable expectation?

To develop a healthy insurance market, insurance companies must deem the market space profitable; that is, the profit gained from underwriting risk cannot be eclipsed by the financial loss of an actual cyber-attack. Boundaries must limit what is within the responsibility of an organization and the insurer and what is beyond both of them. Within that gap area, the federal government must step in as the insurer of last resort, belaying some of the risk. Without federal government participation, the cyber-insurance market cannot expand to meet the safety needs.

Without a flourishing cyber-insurance market, organizations must adopt alternative strategies to mitigate the risk associated with cyber-attacks. CCE aims to fill that gap by providing a scalable cybersecurity framework that can be employed by an individual, organization, or government and customized based on their own risk tolerance.

The "Future" Analysis Problem and Consequence Prioritization

As noted previously, one of the primary challenges in developing secure cyber-systems stems in part from the speed at which adversary capabilities evolve. Organizations are caught in a constant cycle of vulnerability identification and mitigation based on the latest vendor advisories and threat reports. The main issue with this approach, however, is that organizations maintain a reactive posture, responding and mitigating vulnerabilities only after they have been identified. As many have noted, *it is difficult to make predictions, especially about the future.*

This view is echoed by Colbaugh and Glass [24], who note that the "fundamental issues associated with the dynamics and predictability of the coevolutionary 'arms race' between attackers and defenders has yet to be resolved." Although academic efforts have aimed to provide clarity on the means of prediction related to the potential exploitation of a specific vulnerability [24, 25], these pieces are limited in their tendency to assume a correlation between current adversary activity and future capability, or that existing vulnerability scoring systems correlate to the cyber-risk posed to an organization (i.e., the likelihood that a vulnerability will be exploited). In reality, there is a complex system that dictates whether a specific vulnerability will be targeted, one that is based on a variety of factors including existing capability, funding, motivation (e.g., desired end effect), and state-sponsor interest. From an organization's perspective, the challenge to identifying the most significant risk is vast.

In spite of the challenges associated with determining the risk of a cyber-attack, organizations need a method to prioritize resources within a resource-constrained environment. Organizations cannot expect to eliminate the risk (eliminate all of the vulnerabilities); they must therefore identify the means to persist in spite of the risk.

Risk is often defined within the context of the equation:

$$\text{Risk} = \text{Probability} \times \text{Impact}$$

Given this definition, it is possible to describe risk in terms of the potential impact or consequence. If constant probability is assumed, then there is a proportional relationship between risk and impact. Even without an assumption of a constant for probability, the potential impact of an event can still yield significant risk. CCE works within this construct to identify the most significant cyber-events (those with the highest impact) and focus mitigation efforts against those. In doing so, CCE provides an alternative means for prioritization separate from vulnerability scores or the probability of exploitation.

Consider the following hypothetical example. A chemical production facility employs a Pd-based catalytic reaction for the production of vinyl acetate from ethylene, acetic acid (ACOH), and oxygen [26], and its operations can be divided into two parts: reaction and refinement [27].

Information/operations personnel identify a potential vulnerability within the controllers responsible for the distillation column. Although exploitation of this controller by a cyber-adversary could negatively impact the product produced, many of these conditions are recoverable. For example, the product could be refined in another cycle to remove more of the contaminants. Because of this recovery path, the potential impact from this event is relatively low.

In contrast, if those affected controllers were used during the initial reaction, then the potential impact of a cyber-attack is greater. Under normal conditions, the reaction is tightly controlled so that undesirable products are not formed from side reactions, mainly CO_2, the generation of which complicates heat removal from a highly exothermic reaction [27]. In this case, cyber-attacks directed against critical functions of the reaction (rather than refinement) process are more expensive, and therefore the vulnerabilities within controls on this side of the process more significantly represent greater risk.

One significant advantage of the CCE approach and its focus on impact is the ability to merge cybersecurity experience and analysis with a level of engineering expertise that has typically not been included in the conversation. When assessing the technical impact, the subject matter expertise is invaluable for not only determining impact at a single component level but also discovering how that exploitation will impact operations across an infrastructure or region. This expertise exists primarily in the private sector, within the organizations responsible for the day-to-day operations of these systems, highlighting the need for increased and improved communication between the private sector and the federal government.

Although CCE primarily focuses on impact, especially within the early stages of the process, there is value in assessing the probability of a cyber-event occurring. In this case, the mutually beneficial relationship of the federal government and the private sector allows for an exchange of information relevant to assessing probability and impact. The federal government collects and maintains a wealth of threat actor information that is relevant to the cyber-risk mitigation decisions made by various organizations within the private sector. For example, without this input, an organization may choose to mitigate a high-consequence event or scenario that can be accomplished by a cyber-adversary in a relatively small number of actions (e.g., five), rather than a similarly catastrophic scenario that requires multiple actions (e.g., ten), solely on the basis that the first scenario is more likely. Although the second scenario may require multiple actions chained together for success, it is possible that an adversary has already developed seven of the required ten steps, leaving only three remaining. In this case, the probability for the second scenario occurring is arguably greater.

Improved information sharing results in one additional benefit. Within the USA, the majority of critical infrastructure is owned (and operated) by the private sector. It is a stakeholder in any efforts to improve the resiliency or security of critical infrastructure assets. This fact is especially significant in efforts to improve the resiliency of those assets significant to national security. Unfortunately in some cases, the private sector owners of an infrastructure are unaware of the importance of their systems from a national security perspective and therefore do not take that fact into consideration when making risk mitigation decisions.

System of Systems Analysis

If the first stage of CCE focuses on impact, then the second stage of CCE – System of Systems Analysis – focuses on interdependencies. That is, CCE not only considers the cybersecurity of technology but also evaluates the processes and procedures for how the technology is used. CCE also considers the interconnections between these technology boundaries and critical information exchanges. For the organization seeking to develop resilient, but impact-aware architectures, it is also necessary to consider the security of a technology across its entire life cycle.

Following the December 2016 cyber-attack in Ukraine during which a transmission utility was targeted, cyber-researchers identified a malware module that implements a denial of service against Siemens SIPROTEC relays, rendering them unresponsive [28]. The identification of this malware was significant as it highlighted the existence of active adversary activity targeting these systems. At the same time, more recent reporting by Kovacs [29] indicates that even following 2016, vulnerabilities within these systems remain. Still, electric utilities and others are not oblivious to the importance of securing these devices. On the contrary, NERC notes that "Reliable protection systems are necessary for the Bulk Electric System to meet system performance requirements in the NERC Reliability Standards" [30].

The importance of these devices is well known, and access to them or to their settings is generally tightly controlled. These controls are enacted after the equipment is received by the utility and, in some cases, only after a device has been configured, however. A holistic approach to cybersecurity throughout the life cycle of these devices should acknowledge and attempt to control the cyber-risk posed to technology from a cyber-attack that can be introduced via a variety of means, including network-based, human-enabled, and supply chain. For example, although a utility may stringently limit who has access to these devices and who has the appropriate roles and responsibilities for making changes, in many cases the equipment used to make these changes (e.g., a field technician's laptop) are transient assets that cross a variety of security boundaries. The mobility of these devices provides an opportunity for infection or, more significantly, a targeted cyber-attack that takes advantage of an organization's update or validation process.

Cyber-attacks that target an organization's process and procedures are not new; in fact, it is possible to trace this activity to at least 2013, although this adversarial approach may have existed much earlier. In 2013, cybersecurity company F-Secure published its findings regarding Havex, a remote access Trojan that targeted the legitimate update path for some ICS equipment. By targeting these updates, the Havex Trojan eliminated the need to develop a capability to access the victim's systems, instead piggybacking onto normal update activity.

Havex illustrated the danger associated with supply chain targeted attacks. One reason these attacks are so effective is because they target the process and procedures for technology updates, when the traditional security restrictions may be relaxed. For example, energy management systems, critical for situational awareness and control of electric grids, are often considered one of the most tightly controlled systems within a utility. During an update, however, the security boundary of these systems significantly expands, increasing the number of people with access to it and the sensitive business information related to its configuration. Both vendors and manufacturers participate in the update process, greatly increasing the opportunities for an adversary to target this system.

Ultimately, organizations seeking to develop and maintain consequence-aware architectures as a method to improve resilience should conduct a system of systems analysis to understand the interdependencies and chokepoints of their system throughout the entire technology life cycle.

Adversarial Approach and Consequence-Based Targeting

One consistently popular approach to developing resilient architectures is to employ an adversarial approach to the analysis of these systems, processes, and procedures. Often referred to as "red-teaming," it has been described as "the practice of viewing a problem from an adversary or competitor's perspective" [31]. The US Department of Energy's Sandia National Laboratories has adopted a similar perspective for their Information Design Assurance Red Team (IDART) process, noting that this approach is "based on the premise that an analyst who attempts to model an adversary can find systemic vulnerabilities in an information system that would otherwise go undetected" [32]. By adopting an adversarial approach, CCE and other analytic methods redefine the security strategy and mitigation process. Potential cyber-attack scenarios are considered not based on the known adversary activity (e.g., threat reporting, vulnerability disclosures) but on what an adversary must do to be successful.

As noted previously, part of this effort is built around the review of a variety of access methods, including network-based, human-enabled, and supply chain introduction. However, much of this work should also focus on determining the information requirements for a successful attack; where does this information reside and to what information does an adversary need access? In some cases, a review of this type will illuminate where and when an organization's security policy has been relaxed, identifying the spread of information within and beyond the organization.

For example, third-party integrators and ICS vendors often retain business-sensitive information belonging to their customers. With this information, these organizations can provide customized solutions, as well as pointed troubleshooting and technical support. Although the support provided by these third-party groups is invaluable, their assistance comes with the added drawback of potentially increasing their risk of a devastating cyber-attack. This is because implementation vulnerabilities and process exploits greatly outnumber their design counterparts [33]. That is, although there are generic classes of ICS attacks that target design features of a system, specific process attacks remain numerous and more devastating. However, the requirements for enacting these kinds of attacks are also more stringent. An interested adversary needs access to a much greater level of technical detail to develop a feasible cyber concept of operations.

The fact that these complex, catastrophic attacks require such detailed information is a strength of the defender. In most cases, this level of information is tightly controlled, thereby yielding a natural bottleneck or chokepoint for adversary activity that can be closely monitored and serve as an indicator for adversary interest and progress. By conducting an internal review of their systems through CCE, organizations have the opportunity to identify these information troves and assess the degree to which this sensitive information has left the boundary of the organization, serving as a loose indicator of their cyber-exposure.

It should be noted that the purpose of CCE is not to identify those organizations that tightly control their sensitive business information and those who do not. In reality, every organization has some amount of sensitive business information in the open. CCE often employs a period of open source collection to assess what information is publically available on the Internet. This is a useful exercise related to the adversarial approach, as this work represents one of the more common first steps an adversary will take when targeting an organization.

Open source collection represents a relatively low-cost method for collecting the initial organization and process data necessary for targeting. In addition to the relatively low cost, these collection efforts can be fairly covert as well, as a variety of methods exist to obfuscate the online activities of state-sponsored cyber-actors. In contrast, collection on target is a riskier endeavor, with substantially more opportunity for an organization to detect a cyber-actor during their harvest activities.

Open source collection does not eliminate the need for collection on target, however. Inevitably, at some point in the attack development process, a cyber-adversary will require organization-specific information, either as part of a complex cyber-attack on the process or to achieve the deep-level access required.

Mitigation and Elimination of Risk

At its core, the central goal of CCE remains not the complete elimination of cyber-risk, but the introduction of engineering controls that remove the possibility of a cyber-induced catastrophic event. The ease at which these new controls can be introduced varies from system to system, but nearly every case requires the active participation of knowledgeable, experienced subject matter experts. Ultimately, the engineers themselves must introduce the changes to the system, potentially completely reengineering the designs, lessening or eliminating the possibility for a cyber-based attack against that function or service.

For example, consider a cyber-based attack directed against variable frequency drives in which a cyber-adversary, seeking to cause physical damage, sets these drives to operate at a resonance frequency (skip frequency) [34]. By design, engineers program a bypass, skipping the skip frequency to ensure that no damage comes to the device. From an adversary perspective, the opposite can also be true. That is, by using the engineering documentation, an adversary can identify a damaging frequency and set the variable frequency drives to operate at that resonance. One mitigation strategy may be the installation of an independent input/output device that gracefully shuts down the drives as they approach resonance frequencies.

In some cases, the introduction of security controls requires the active participation of the vendors and manufacturers of ICS software and hardware, especially when the vulnerabilities identified curing CCE are design in nature. In these cases, their elimination may require a complete rework of a product, a potentially expensive proposition from a resource perspective. The degree to which these more extreme solutions should be enacted is an analytic challenge in of itself and one that may require participation by the federal government or other large collective body. In many cases, a single organization may not have the leverage necessary to request alterations of this magnitude; in contrast, a consortium of multiple organizations may provide the incentive to induce the reform of ICS equipment.

In the case that an identified scenario poses a risk to an asset critical to national security, then a role exists for participation by the federal government.

There is another opportunity for engagement from the federal government, however, primarily in the development of suitable trip wires for adversary capability development monitoring. The federal government has established a large and complex system for the collection and processing of cyber-threat actor information. Still, there exists the opportunity for improvement in the sharing of this information. CCE is most effective, as an iterative process, with organizations completing multiple CCE studies in series to lessen the most significant cyber-impacts. As an organization completes these studies, their findings can be used to improve intelligence collection efforts over time, while the improved threat data can ensure that organizations continue to expend their limited resources where they can be most impactful.

References

1. R. Lee, M. Assante, T. Conway, Analysis of the cyber-attack on the Ukrainian power grid. Prepared for the Energy Information Sharing and Analysis Center (E-ISAC), 16 Mar 2016
2. S. Gorman, Electricity grid in U.S. penetrated by spies. (Wall Street J, 2009), https://www.wsj.com/articles/SB123914805204099085

3. S. Toppa, The National Power Grid is under almost continuous attack, report says. (Time, 2015), http://time.com/3757513/electricity-power-grid-attack-energy-security/
4. N. Perlroth, Hackers are targeting nuclear facilities, Homeland Security Dept. and F.B.I. say. (The New York Times, 2017), https://www.nytimes.com/2017/07/06/technology/nuclear-plant-hack-report.html
5. L. Dearden, Russian cyber-attacks have targeted UK energy, communication, and media networks, says top security chief. (Independent, 2017), https://www.independent.co.uk/news/uk/home-news/russia-hacking-uk-bt-media-energy-companies-top-spy-security-schief-a8055371.html
6. J. Borger, US accuses Russia of cyber-attack on energy sector and imposes new sanctions. (The Guardian, 2018), https://www.theguardian.com/us-news/2018/mar/15/russia-sanctions-energy-sector-cyber-attack-us-election-interference
7. S. Karnouskos, Stuxnet worm impact on industrial cyber-physical system security, in *IECON 2011-37th Annual Conference on IEEE Industrial Electronics Society*, 2011
8. D.P. Fidler, Was Stuxnet an act of war? Decoding a cyberattack. IEEE Security & Privacy **9**(4), 56–59 (2011)
9. A. Matrosov, E. Rodionov, D. Harley, J. Malcho, Stuxnet under the microscope. ESET, Technical report, 2011, revision 1.31
10. M. Braglia, MAFMA: multi-attribute failure mode analysis. Int. J. Qual. Reliab. Manag. **17**(9), 1017–1033 (2000)
11. A. Bolshev, J. Larsen, M. Krotofil, R. Wightman, A rising tide: design exploits in industrial control systems. Prepared for 10th USENIX workshop on offensive technologies, WOOT 16, USENIX Association, Austin, TX, 2016
12. Sandia National Laboratories, Guide to CIP cyber-vulnerability assessment, http://energy.sandia.gov/wp-content/gallery/uploads/CIP_CyberAssessmentGuide.pdf
13. C. Ten, C. Liu, G. Manimaran, Vulnerability assessment of cybersecurity for SCADA systems. IEEE Trans. Power Syst. **23**(4), 1836–1846 (2008)
14. P.A.S. Ralston, J.H. Graham, J.L. Hieb, Cybersecurity risk assessment for SCADA and DCS networks. ISA Trans. **46**, 583–594 (2007)
15. Symantec, Internet security threat report. (2016), https://www.symantec.com/content/dam/symantec/docs/reports/istr-21-2016-en.pdf
16. M. Dacier, L. Yumer, T. Dumitras, Lessons learned from a rigorous analysis of two years of zero-day attacks. Prepared for RSA conference Asia Pacific, 2013, https://www.rsaconference.com/writable/presentations/file_upload/cle-t02_final_v2.pdf
17. L. Ablon, A. Bogart, Zero days, thousands of nights: the life and times of zero-day vulnerabilities and their exploits. (Rand, 2017), https://www.rand.org/content/dam/rand/pubs/research_reports/RR1700/RR1751/RAND_RR1751.pdf
18. S. Tom, D. Christiansen, D. Berrett, *Recommended Practice for Patch Management of Control Systems* (Department of Homeland Security, Washington, D.C., 2008). https://ics-cert.us-cert.gov/sites/default/files/recommended_practices/RP_Patch_Management_S508C.pdf
19. C. St Michel, S. Freeman, R. Smith, M. Assante, Consequence-driven. (Cyber-Informed Engineering. 2016), https://www.osti.gov/biblio/1341416
20. R. Pal, L. Golubchik, K. Psounis, P. Hui, Security pricing as enabler of cyber-insurance *A First Look at Differentiated Pricing Markets*. IEEE Trans. Dependable Secure Comput. (2016)
21. N.S. Malik, R. Collins, M. Vamburkar, Cyberattack pings data systems of at least four gas networks. (Bloomberg, 2018), https://www.bloomberg.com/news/articles/2018-04-03/day-after-cyber-attack-a-third-gas-pipeline-data-system-shuts
22. S. Romanosky, L. Ablon, A. Kuehn, T. Jones, Content analysis of cyber-insurance policies: how do carriers write policies and price cyber-risk? (Rand Corporation, 2017), https://ssrn.com/abstract=2929137
23. M. Thompson, Why cyber-insurance will be the next big thing. (CNBC, 2014), https://www.cnbc.com/2014/07/01/why-cyber-insurance-will-be-the-next-big-thing.html

24. R. Colbaugh, K. Glass, Proactive defense for evolving cyber-threats, in *IEEE International Conference on Intelligence and Security Informatics*, Beijing, China, 2011, https://www.osti.gov/servlets/purl/1108387

25. M. Bozorgi, L. Saul, S. Savage, G. Voelker, Beyond heuristics: learning to classify vulnerabilities and predict exploits, in *Proceedings of the 16th International Conference on Knowledge Discovery and Data Mining*, 2010, pp. 105–114

26. Y.-F. Han, D. Kumar, C. Sivadinarayana, D.W. Goodman, Kinetics of ethylene combustion in the synthesis of vinyl acetate over a PD/SiO2 catalyst. J. Catal. **224**, 60–68 (2004)

27. D. Gollmann, P. Gurikov, A. Isakov, M. Krotofil, J. Larsen, A. Winnicki, Cyber-physical systems security – experimental analysis of a vinyl acetate monomer plant. ACM Cyber-Physical System Security Workshop (CPSS), Singapore, 2015

28. A. Cherepanov, Win32/Industroyer: a new threat for industrial control systems. (ESET, 2017), https://www.welivesecurity.com/wp-content/uploads/2017/06/Win32_Industroyer.pdf

29. E. Kovacs, Electrical substations exposed to attacks by flaws in Siemens devices. (2018), https://www.securityweek.com/electrical-substations-exposed-attacks-flaws-siemens-devices

30. Havex hunts for ICS/SCADA systems. (F-Secure Labs, 2014), https://www.f-secure.com/weblog/archives/00002718.html

31. R. Heuer Jr., R. Pherson, *Structured Analytic Techniques for Intelligence Analysis* (Sage/CQPress, Washington, D.C., 2015)

32. B. Wood, R. Duggan, Red teaming of advanced information assurance concepts, in *DISCEX 2000*, Hilton Head, South Carolina, January 2000, http://cs.uccs.edu/~cchow/pub/master/sjelinek/doc/research/red.pdf

33. J. Larsen, Physical damage 101: bread and butter attacks. (Blackhat, 2015), https://www.blackhat.com/docs/us-15/materials/us-15-Larsen-Remote-Physical-Damage-101-Bread-And-Butter-Attacks.pdf

34. R. Wightman, The easy button for cyber/physical ICS attacks, in *S4 Security Conference*, 2016

Part II
Cyber-Modeling, Detection, and Forensics

Cyber-Physical Anomaly Detection for Power Grid with Machine Learning

Pengyuan Wang and Manimaran Govindarasu

Abstract The power system is one of the most critical infrastructures in modern society. As a sophisticated cyber-physical system (CPS), its operation highly relies on the tight coupling between the physical layer (electric energy carrier) and the cyber layer (data and information carrier). To maintain high availability and security of both cyber and physical layer is critical in order to guarantee that the electricity generation and consumption process is not disturbed. However, various factors, such as natural device defects, human mistakes, and malicious cyber-activities, can all result in severe interruption of the operation of a power system, among which cyber-sabotage is the most unpredictable and uncontrollable.

To improve the situational awareness and hence help enhance the cybersecurity of the modern power grid, cyber-physical anomaly detection that leverages machine-learning techniques is investigated in this chapter. Firstly, a brief overview about the modern power systems and their cybersecurity concerns is given. Then, facets of power grid anomaly detection and suitable applications of various machine learning techniques are elaborated. Finally, we select generation control anomaly detection as a study case to provide more insights.

Cybersecurity of Modern Power System

Overview of Modern Power System

Modern power systems have grown into sophisticated cyber-physical systems due to the expansion of their electrical infrastructures and the consequential application of diverse communication and information. A modern power system comprises two layers that are tightly coupled: the energy (physical) layer and the information (cyber) layer. The physical layer carries the electricity from end to end and is the

P. Wang · M. Govindarasu (✉)
Department of Electrical and Computer Engineering, Iowa State University, Ames, IA, USA
e-mail: pywang@iastate.edu; gmani@iastate.edu

© Springer Nature Switzerland AG 2019
C. Rieger et al. (eds.), *Industrial Control Systems Security and Resiliency*, Advances in Information Security 75, https://doi.org/10.1007/978-3-030-18214-4_3

core of the power system; however, the information layer is of the same importance since it acts as the nervous system of a power grid.

The physical layer of a power system contains four major domains: generation, transmission, distribution, and consumption [1]. For the last two decades, power systems have become more heterogeneous in terms of all these domains. Renewable energy plants, flexible AC transmission system (FACTS), high-voltage direct current (HVDC) transmission, distributed energy resources (DER), and micro-grids, electricity storage, electric vehicles, etc. have all been gradually integrated into the traditional bulk power system. Their emergence brings great challenges for the operation of the power system because the coordination of different system components becomes even more complex. As a result, more efficient control schemes are incorporated in the cyber-layer. The cyber-layer of a power system is composed of a large amount of secondary devices and schemes used to monitor, protect, and control the operation of the energy layer. Secondary devices in the cyber-layer are capable of communication, data collection, data storage, data processing, decision-making, control actuation initialization, etc. They can be intelligent electronic devices (such as phasor measurement units [PMUs], smart meters, and relays), or more powerful hardware that accommodate commercial software running at a regional market or control center. Components in the cyber-layer can be interconnected via public or dedicated communication networks. The geographic scope of the communication varies from case to case, however, wide-area communication has become a prerequisite. For a modern power system, the three most critical cyber-layer systems include the supervisory control and data acquisition (SCADA), the wide-area measurement system (WAMS), and the advanced metering infrastructure (AMI).

Cyber-Systems Relying on Wide-Area Communication

SCADA (Supervisory Control and Data Acquisition)

A SCADA is widely used for industrial automation system monitoring and control. In a power system, each station has a remote terminal unit (RTU) that merges data from local secondary devices such as transducers, sensors, and meters. SCADA collects data from RTUs located at different stations and forwards them to the control center via wide-area communication every few seconds. Applications in energy management systems (EMSs), including state estimation (SE), automatic generation control (AGC), economic dispatch, contingency analysis, wide-area protection, and control schemes, utilize those data to monitor the operational status of the power system and send control commands back to the stations accordingly. The communication protocol commonly used in SCADA systems by North American utilities is DNP3, and it still lacks security features such as encryption or authentication.

WAMS (Wide Area Measurement System)

With the advent of phasor measurement units (PMUs), the wide area measurement system (WAMS) emerges as an alternative to SCADA, which collects data at a much higher sampling rate. The typical sample rate adopted in WAMS is 30, 60, or 120 samples/s. With time-synchronized PMUs, WAMS has access not only to voltage or current magnitudes but can also measure relative phasor angles directly. The WAMS is often constructed in a hierarchical architecture, where phasor data concentrators (PDCs) at different levels will first concentrate the data from PMU and then forward the processed data to the control center. Nowadays, practical applications based on WAMS are mainly for monitoring, such as transient instability detection and linear state estimation. Wide-area protection and controls based on synchrophasors are also under development. More and more applications will be put into use when the data accuracy of WAMS gets further improved. For WAMS, it often adopts IEEE C37.118 as the communication protocol.

AMI (Advanced Metering Infrastructure)

Unlike SCADA and WAMS, advanced metering infrastructure (AMI) is a system placed at the distribution sector that monitors, collects, and analyzes the energy usage data of consumers. Real-time energy usage data are collected via a large number of smart meters, and proper demand side controls can be derived to improve the operational reliability of the bulk power system. The controls could be achieved through adjustments of the electricity price. The two-way communication is composed of the wide-area network, the neighborhood-area network, and the home-area network. The commonly utilized communication protocol is ANSI C12 series in the USA and IEC 62056 in the EU [2].

Greenhouse for Malicious Cyber-Attacks

The heavy deployment of communication and information technology (IT) facilitates the monitoring and control of the modern power system. However, one side effect is barely considered in old-time implementations—security issues of the cyber-layer. Though it is a popular topic in computer science, it was not getting enough attention when the critical infrastructures were constructed. The main goal of power system designers was to enable the data exchange and realize the monitoring/protection/control functions and how secure those communications really are was often ignored.

Due to the homogeneous nature of any information networks, nothing really stops malicious cyber attackers from targeting the power grids. Recent reported attacks involve trials to compromise utility corporate networks, control centers, wide-area protection and control schemes, field devices, etc. Any successful attacks, especially those trying to sabotage the operation of bulk power system (BPS), can cause severe loss. The application of conventional IT security technologies is necessary, but they are often insufficient to mitigate sophisticated attacks targeting CPS, which deploy zero-day exploits or social engineering tactics. PJM Interconnection's former CEO, Terry Boston, once mentioned in a report that "the utility experiences 3,000–4,000 hacking attempts every month" [3]. Around 2010, the Stuxnet malware that specifically compromises ICS PLCs became well known and demonstrated a high level of stealth and sophistication to spoof the telemetry information deceiving the operators for a long period of time [4]. On December 23, 2015, a cyber-attack took place in the Ukraine power grid after a 6-month-long reconnaissance, where the attackers successfully disconnected 30 distribution substations for about 3 hours, causing significant outages to thousands of customers [5]. As can be seen from these examples, the security issues of power systems are more severe than people may believe. Cyber-threats, vulnerabilities, and the malicious exploits of both can induce huge impacts on the normal operation of a power system.

In order to better secure cyber-physical systems, efforts could be made in three different ways: (1) prevention, (2) detection, and (3) mitigation. Prevention represents the type of countermeasures which recognize any potential vulnerabilities and cyber-threats before they get exploited and carry out preventive actions to dispose. Detection is utilized to detect ongoing anomalous behaviors in real time, while mitigation is the last resort to constrain the impactful propagation of the attacks and enhance power system resilience. From the description of each of these defense methods, we can see that detection plays a key role in system operation monitoring and helps improve the situation awareness and lays the foundation for both prevention and mitigation. Development and application of anomaly detection are popular in the conventional IT security sector and play a critical role in protecting the data integrity and privacy. But the mature IT anomaly detection technologies cannot solve all the problems in a cyber-physical system. For a CPS, it has unique features and requirements that need to be considered and handled by a proper CPS anomaly detection module.

Cyber-Physical System Anomaly Detection

Anomaly detection is the basic capability that the critical CPS functionalities should possess in order to be situation-aware and further carry out adaptive preventive or corrective defenses. For a specific system, anomalies refer to the behavior patterns that exhibit distinctions from the normal ones. The task of anomaly detection for a CPS is to detect the anomalies accurately and timely by collecting and analyzing the data from both cyber and physical layers.

CPS Anomaly Detection Applications

The application of anomaly detection in modern power systems is widely investigated due to the enormous emerging cyber-threats. In this section, two different taxonomies are presented for the application of anomaly detection. First is the location-based taxonomy as shown in Fig. 1, while the second taxonomy is according to the nature of the detection.

Detection Location

1. *Centralized anomaly detection*—Anomaly detection often requires global system features. Thus, it is natural to place it at the location where the global data is easily accessible, for instance, in the control center. Anomaly detection in a control center (or even lower in the hierarchy, as is shown in Fig. 1) can utilize global physical measurements as detection features. Work that has been observed includes the following:

   - PMU bad data detection [6].
   - Bad measurement detection for critical EMS applications [7].
   - General anomalous power system behavior detection [7].

   Centralized anomaly detection implemented at the control centers concentrate on global data and information, such that more accurate decisions about the anomaly can be obtained. However, as a centralized module, it is vulnerable to cyber-attacks as well and could become the single point of failure.

2. *Network anomaly detection*—Conventional cyber-intrusion detection systems (IDS) can be applied in a networking environment. By mainly leveraging the networking features, IDS are able to detect many malicious cyber-activities. For example, a large amount of ICMP packets being received at a host could be a symptom of host-scanning or denial-of-service (DoS) attacks with ping flood. From the perspective of monitoring scope, an IDS can be host-based or network-based.

3. *Field anomaly detection*—Centralized anomaly detection brings excellent performance. However, centralized anomaly detection cannot do much regarding any malicious attack trials along the actuation path. For example, when the attacker launches a data integrity attack on the AGC control commands, the centralized anomaly detection algorithm will not be able to detect. Therefore, another possible CPS anomaly detection is field detection. It deploys the detection at the field layer where the actuations are carried out but can help detect anomalies either induced by bad measurements or manipulated control commands.

   The anomaly detection in the field layer can be host-based. For instance, it can be equipped for each intelligent electronic device (IED), and the anomaly is defined within the scope of an IED operation. However, the host-based anomaly

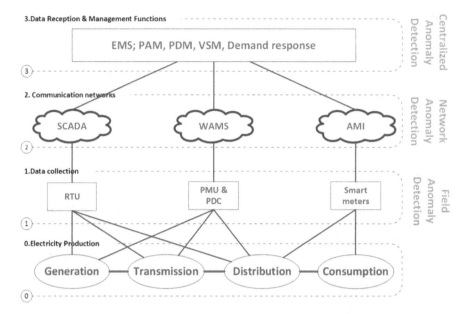

Fig. 1 Abstract architecture of CPS and possible locations of anomaly detection

detection is still limited due to lack of the wider global view and is likely to miss the coordination of simultaneous events. One alternative to host-based field anomaly detection is the decentralized anomaly detection architecture. By enabling the information and data exchanging among multiple stations and driven by the large volume of data accessible, the field stations can work in a cooperative manner for global anomaly detection.

Detection Nature

According to the detection principle adopted, an IDS can be signature-based, anomaly-based, specification/model-based, or data-driven-based.

1. *Signature-based*—Signature-based IDS checks the predefined signatures or patterns of malicious cyber-behaviors in order to detect known malwares or attacks. Different detection rules need to be predesigned manually based on empirical knowledge, which is a knowledge-intensive process. This is often cost-prohibitive for a large-scale system. Another disadvantage of signature-based detection is that it cannot identify unknown attacking patterns.

2. *Anomaly-based*—Unlike signature-based anomaly detection, anomaly-based IDS utilizes statistical methodologies to find behaviors that deviate from normal ones. But problems occur when the limited statistics considered cannot fully distinguish the new types of attacks.

3. *Specification/model-based*—Researchers have proposed a specification-based anomaly detection to bring the stateful understanding of the functionalities [2] and better distinguish the new anomalies from the normal system states. The main drawback of this methodology is that its scalability is constrained by the logical simplicity of the functionality. If the logics of the function is too complex to specify, then this method cannot be easily applied.

Other model-based anomaly detection [7] is also frequently used. By having a predetermined model of the power system, the expected behavior is predicted beforehand and will be leveraged as a reference for real system behavior. The shortcoming here is that the model cannot receive online updates after its installation, which makes detections inaccurate if the model becomes outdated.

4. *Data-driven-based*—Machine-learning techniques have provided a powerful tool for CPS anomaly detection. Similar to anomaly-based detection, data-driven detection compares the online data patterns exhibited with the normal ones to detection anomalies. But instead of utilizing simple statistics, it can incorporate a great amount of features into the detection, while it still has an efficient way to handle the data-mining process. Table 1 summarizes the technologies that can be applied in CPS anomaly detection.

Table 1 Anomaly detection techniques

Type		Method	Nature
Conventional methods		Signature-based Anomaly statistics-based Model/specification-based	Empirical or statistical methods
Machine-learning methods	Classification-based	Support vector machine Decision tree Random forest Bagging AdaBoost XGBoost Neural network Other ensemble methods	Supervised learning
	Clustering-based	K-means clustering Hierarchical clustering Density-based clustering	Semi-supervised learning Unsupervised learning
	Dimensionality reduction	Domain knowledge Principal component analysis (PCA) Non-negative matrix factorization (NMF) Kernel PCA ISOMap T-SNE	Data pre-processing

Challenges for CPS Anomaly Detection

Timing Performance

Unlike enterprise network cyber-anomaly detection, the operation of critical infrastructures often has stringent timing requirements for the availability of relevant
functionalities. If the anomaly detection algorithm is able to detect the anomaly, but
takes too long to prevent the anomaly from causing de facto impacts, it is below
expectations since the main impetus for people to adopt anomaly detection is to get
prepared with prevention actions in time.

In a power system, the functional operation time granularity varies greatly. For
example, the allowed operation time for wide-area protection schemes ranges from
hundreds of milliseconds to a few minutes, automatic generation control (AGC)
operates approximately every 4–8 seconds, and economic dispatch runs every
5 minutes. Thus, when developing the anomaly detection system, it is important to
consider the timing performance of the algorithms beforehand, according to the
functionality it protects.

Big Data

For data-driven anomaly detection, the processing of big data often becomes a
challenge. Volume, variety, velocity, and veracity are the four facets of the data
that should be considered when trying to develop a data-driven anomaly detection
system:

- Volume—Like many other industrial critical infrastructures, power systems
  operate continuously 24/7; thus, the sample measurements easily accumulate.
  For example, the 1-day second-level power consumption data of all homes in
  New York state could be 127.1 TB [8].
- Variety—Sophistication of power systems have determined the variety of the
  data. Data from sources such as transducers, IEDs, control centers, and networking devices can be in the form of nominal data, numeric data, log texts, and even
  video/audio files. Reference [9], has provided a list of possible data sources in the
  smart grid.
- Velocity—The sampled data, such as SCADA and phasor measurements, are fed
  into the control center as data streams. For the phasor measurements specifically,
  its update speed will be quite fast. The time interval between the two phasor
  samples could be as short as 10 milliseconds.
- Veracity—Due to the flaws inherent in the sensing devices and data transmission
  paths, the measurements could be incorrect. Therefore, bad data should also be
  treated as a possible type of anomaly.

From the perspective of the data types, data that can be utilized in anomaly detection systems include raw physical measurements, processed physical measurements, and cyber-measurements, as follows:

- Raw physical measurements—Breaker/switch status, bus voltage phasor, bus power injection, line active/reactive power; generator terminal frequency and system frequency, etc.
- Processed physical measurements—Metrics derived from raw physical measurements that contain dense information. For example, the time interval between the two sequential frequency dips may reveal the tendency of the cascading events.
- Cyber-measurements—Much information can also be obtained from the cyber-layer devices. The characterization of communication protocols, packet types/patterns, routing device logs, etc. can all be utilized as attack indicators.

Considering that the data availability is limited to the communication network topology, the data accessible to one specific anomaly detection system could also be classified as:

- Public local information—Local information that could be shared with other functions on request.
- Private local information—Sensitive local data that will not be shared with other functions at different locations.
- Global information—Information that has been successfully collected from the community.

Detection Model Online Update

Measurement data from either the SCADA, WAMS, or AMI are streaming into the anomaly detection modules. New data might contain new events that have never been observed before. This will pose the challenge to update the detection model online. Otherwise, detection accuracy can be reduced due to the outdated detection models.

CPS Anomaly Detection with Machine Learning

In data-driven CPS anomaly detection, the data complexity requires an effective data processing and anomaly detection algorithm. Machine-learning techniques have demonstrated great competence in extracting information from large amounts of raw data. Table 1 previously summarized the common machine-learning techniques for CPS anomaly detection, which are:

Classification Methods When the number of obtained data instances is reasonably small, they could be labeled by experts with the class that they belong to. Then, supervised learning-based classification can be utilized in anomaly detection training. After the training, a "model" can be derived and used for online detection. Many different classification methodologies exist for this purpose. For instance, a decision tree (DT) delivers the detection model as a tree with sequential decision-making logic embedded, while the model that support vector machines (SVM) attain is a hyperplane specified by a group of optimized hyperplane parameters. In the detection stage, a new data instance will be classified according to its location in terms of the hyperplane. Ensemble methods, such as random forest, boosting, that leverage multiple simple classifiers, can often get better overall performance.

Clustering When the amount of data instances is too large to label them all, semi-supervised or unsupervised machine-learning techniques can be applied. Various clustering techniques are listed in Table 1. K-means clustering and hierarchical clustering are distance-based methods that are proper for scenarios with evenly distributed data instances, and density-based methods such as density-based spatial clustering of application with noise (DBSCAN) performs better when the clusters possess different densities.

Data Processing Another key technique in the application of machine learning is dimensionality reduction. For datasets with a variety of features, dimensionality reduction is necessary in order to improve computational efficiency and virtualize the data. How to select the dimensionality reduction technique depends on each specific problem. For some problems, linear techniques such as PCA can perform well, but for the others, if the instances are concentrated on nonlinear manifolds in the original hyperspace, nonlinear dimensionality reduction algorithms, such as t-SNE, are better options. In some special scenarios, if the relationship among features is well known with domain knowledge, it would be better for the domain experts to reduce the dimension by applying their expertise first, which often leads to better results.

Different machine-learning methods will have different timing performance and detection accuracies, and no single methodology can outperform others in any cases. Therefore, the selection of methodologies should be carried out based on the characteristics of each concrete problem.

Case Study: Clustering-Based Generation Control Anomaly Detection

One of our previous work was selected as a case study in order to demonstrate the application of domain knowledge-based dimensionality reduction and clustering in CPS anomaly detection [10].

Problem

The generation and load balance is the most critical requirement for the operation of the power system, which is maintained by various generation control loops— primary generation control (governor action), automatic generation control (AGC), etc. Primary generation control helps maintain system stability at the expense of a small-frequency deviation, and AGC control is able to correct the deviation. AGC is installed for a region called balancing authority (BA); as such, each BA needs to balance its regional generation and load. As shown in Fig. 2, AGC operates by sending an area control error (ACE) to secondary generators, so that ACE is the adjustment that one generator needs to make to recover the balance between system generation and the load.

Cyber-attacks targeting power system generation control can be conducted through the manipulation of generator setting points after intrusion, data integrity attack, false data injection, etc. If the generation control is attacked, the generation is driven off the optimal generation plan, and the attackers might benefit economically through this. On the other hand, the attacker could also maliciously break the generation and load balance to cause load loss or system instability. Therefore, anomaly detection is needed to detect any potential attacks in time. The main idea behind the anomaly detection we have proposed is to propagate the data among the generators within one BA and correlate their behaviors. Theoretically, the generators in the same region should have similar generation adjustment behaviors. If not, it

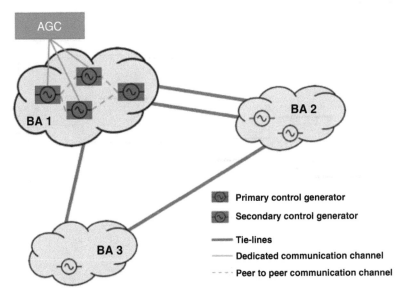

Fig. 2 The organization of the generators in a typical power system [10]

could potentially be an anomaly. Anomaly detection is formulated as a multi-class classification problem, and its task is to distinguish between normal system events and attacks.

Experiment

IEEE 39-bus power system is divided into three BAs, as is shown in Fig. 3, and each BA has its own AGC function. Area 2 is the region under consideration. G5 on bus 34 and G6 on bus 35 are both secondary generators, and in all experiments, G5 will be utilized as the attack target, while G6 functions as its neighbor generator.

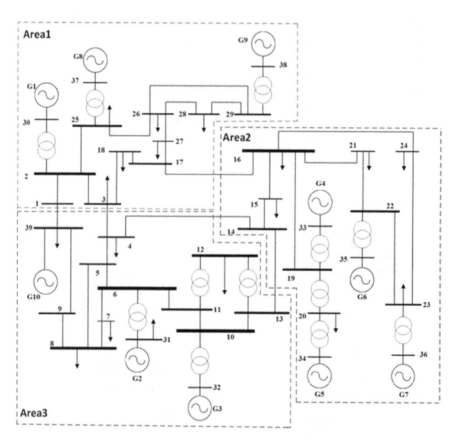

Fig. 3 IEEE 39-bus power system divided into three areas [10]

Scenarios and Data Collection

Six scenarios are simulated to generate the synthetic datasets, including two normal scenarios and four scenarios where the system was under attack.

1. *Normal load change inside the balancing authority.*

    This scenario includes random load changes occurring in the balancing authority of the generators under consideration. When load changes occur in the same BA, secondary generators are expected to adjust their generations to meet the load changes. Figure 4a shows the system frequency and generation levels of G5 and G6 along the simulation time.

2. *Normal load change outside the balancing authority.*

    Random load changes occur outside of the BA of the generators under consideration. When this happens, primary control of generators in this BA can help temporarily, but the disturbance to the generators is not as noticeable as the first scenario since no significant ACE control value should be received. Figure 4b shows the system frequency and generation levels of G5 and G6 along the simulation time, while the generation deviation magnitude is smaller than that in the first scenario.

3. *ACE control under constant value attack.*

    In this scenario, the attacker modifies the ACE control values sent to G5 to a constant value by launching a man-in-the-middle (MITM) attack. The system response is depicted in Fig. 5a. By sending a fake, constant negative ACE value to G5, the attacker ramps up G5 generation level gradually by manipulating the secondary generation control loop. Since the values being sent from the ACE control to G6 is uncompromised, it will try to correct the impacts of G5.

4. *ACE control under flip attack.*

    The attacker changes the sign of every ACE control value sent to G5. Figure 5b shows the attack impacts. No huge disturbances are caused for the system, but the adversary may attack in such a way to obtain a benefit from the market.

5. *Generation ramp attack.*

    Via intrusion into a power plant, the attacker can directly change the generation level of a generator. This case shows that the attacker gets into a power plant of G5 and then quickly ramps it up. The system frequency has a huge deviation due to this attack as seen in Fig. 6a.

6. *Generation switching attack.*

    In this case, as presented in Fig. 6b, the output of G5 causes it to ramp up and down frequently. This type of attack can cause undesirable system oscillation, which may cause severe stability issues.

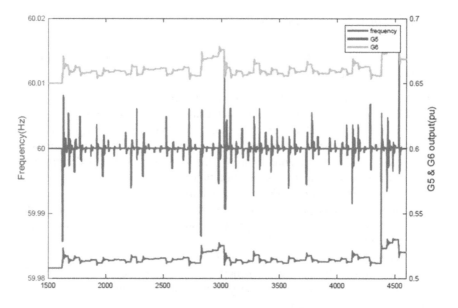

(a) Normal load change inside BA.

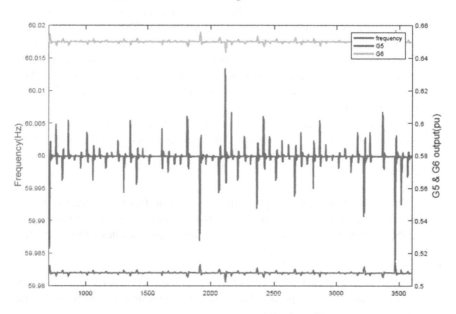

(b) Normal load change outside BA.

Fig. 4 Normal load change. (**a**) Normal load change inside BA. (**b**) Normal load change outside BA

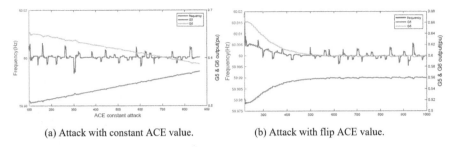

(a) Attack with constant ACE value.   (b) Attack with flip ACE value.

Fig. 5 Attack on ACE value. (**a**) Attack with constant ACE value. (**b**) Attack with flip ACE value

Dimensionality Reduction

Three conformity metrics are defined by Wang et al. [10] based on the domain knowledge to preprocess the raw dataset, and every instance is a point in three-dimensional space. The training instances shown in Fig. 7a are labeled, while those in Fig. 7b depict all of the training instances.

Clustering Results

K-means clustering is run on all of the training instances. Instead of using random centroids for the six clusters, all of the labeled training instances will be used for centroid initialization. Figure 8 provides the final clustering results with all of the training instances. We can see that the clustering is not separating the normal load changes inside the BA and the normal load changes outside the BA; thus, it also mixes the instances under flip attack and switching attack. Online testing is done with a new dataset, and the detection confusion matrix is given as shown Table 2.

The clustering and detection results show that the method proposed has difficulty in distinguishing between the two normal scenarios and the two specific attack scenarios, but it can still successfully distinguish the normal patterns and abnormal ones, which is the key requirement for an anomaly detection algorithm.

Conclusion

The operation of modern power systems significantly relies on the wide-area communication and information technology, and thus the critical monitoring and protective and control functionalities become extremely vulnerable to the cyber-threats.

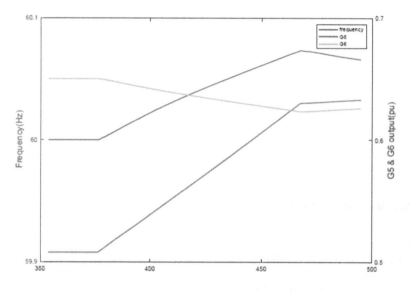

(a) Generation ramp attack.

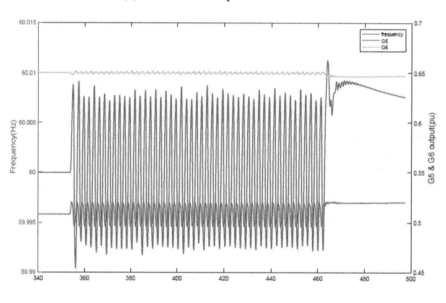

(b) Generation switching attack.

Fig. 6 Generation attack after intrusion. (**a**) Generation ramp attack. (**b**) Generation switching attack

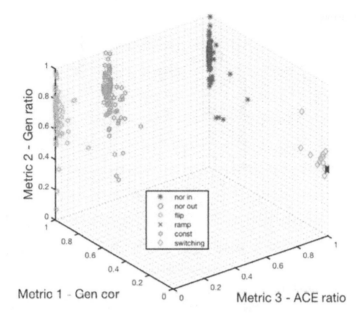

(a) Labeled training instances.

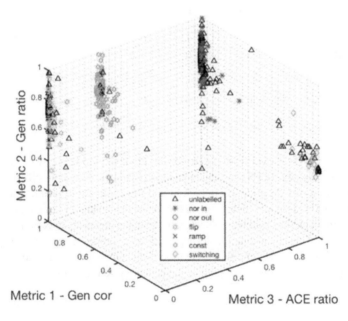

(b) All training instances.

Fig. 7 Training dataset [10]. (**a**) Labeled training instances. (**b**) All training instances

Fig. 8 Clustering results [10]

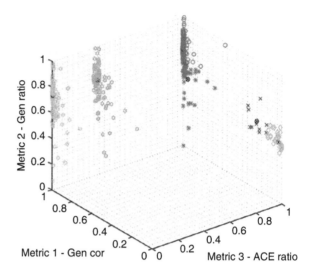

Table 2 Test confusion matrix

	Normal inside	Normal outside	Flip attack	Constant ACE attack	Ramp attack	Switching attack
Normal inside	31.51	67.12	0	1.37	0	0
Normal outside	0	100	0	0	0	0
Flip attack	0	0	100	0	0	0
Constant ACE attack	0	0	0	100	0	0
Ramp attack	0	0	0	0	0	100
Switching attack	0	0	0	0	6.25	93.75

In this chapter, a brief overview is carried out about the modern power system as a cyber-physical system (CPS) and its cybersecurity issues. The main efforts focus on the introduction of CPS anomaly detection and the application of machine-learning techniques. CPS anomaly detection taxonomies are made in terms of detection location and detection nature, respectively, and the general detection challenges are discussed. Classification, clustering, and data preprocessing techniques are briefly introduced as CPS anomaly detection tools. One study case is also provided where K-means clustering is adopted in the power system generation control anomaly detection.

References

1. NIST framework and roadmap for smart grid interoperability standards. [Online]. Available: https://www.nist.gov/sites/default/files/documents/public_affairs/releases/smartgrid_interoperability_final.pdf
2. R. Berthier, W. H. Sanders, Specification-based intrusion detection for advanced metering infrastructures, in *Proc. IEEE Pacific Rim Int. Symp. Dependable Comput. PRDC*, pp. 184–193, 2011
3. J. Dougherty, Biggest U.S. power grid operator suffers thousands of attempted cyber-attacks per month, [Online]. Available: https://forwardobserver.com/2017/08/biggest-u-s-power-grid-operator-suffers-thousands-of-attempted-cyber-attacks-per-month/
4. N. Falliere, L. O. Murchu, E. Chien, W32.stuxnet dossier, Feb 2011. [Online]. Available: www.symantec.com/content/en/us/enterprise/media/security_response/whitepapers/w32_stuxnet_dossier.pdf
5. R. M. Lee, M. J. Assante, T. Conway, Analysis of the cyber-attack on the Ukrainian power grid, in *Electricity Information Sharing and Analysis Center (E-ISAC)*: Washington, DC, USA, 18 March 2016
6. M. Wu, S. Member, L. Xie, S. Member, Online detection of low-quality synchrophasor measurements: A data-driven approach. IEEE Trans. Power Syst. **32**(4), 2817–2827 (2016)
7. S. Sridhar, M. Govindarasu, Model-based attack detection and mitigation for automatic generation control. IEEE Trans. Smart Grid **5**(2), 580–591 (2014)
8. Z. Huang, H. Luo, D. Skoda, T. Zhu, Y. Gu, E-sketch: Gathering large-scale energy consumption data based on consumption patterns, in *Proc. of 2014 IEEE Int. Conf. Big Data, IEEE Big Data* 2014, pp. 656–665, 2015
9. J. Hu, A.V. Vasilakos, S. Member, Energy big data analytics and security: Challenges and opportunities. IEEE Trans. Smart Grid **7**(5), 2423–2436 (2016)
10. P. Wang, M. Govindarasu, A. Ashok, S. Sridhar, D. McKinnon, Data-Driven Anomaly Detection for Power System Generation Control, in *2017 IEEE Int. Conf. Data Mining Work.*, pp. 1082–1089, 2017

Toward the Science of Industrial Control Systems Security and Resiliency

Mohammad Ashiqur Rahman and Ehab Al-Shaer

Abstract The supervisory control and data acquisition (SCADA) system is the major industrial control system (ICS), which is responsible for collecting data from end devices, analyzing data, and managing the system efficiently by sending necessary control commands to the corresponding end devices. Unlike traditional cyber networks, a SCADA system consists of heterogeneous devices that communicate with one another under various communication protocols, physical media, and security properties. Failures or attacks on such networks have the potential of data unavailability and false data injection causing incorrect system estimations and control decisions leading to non-optimal management or critical damages of the system. This chapter provides a theoretical baseline for assessing the security and resiliency of ICS by presenting two formal frameworks, one for security analysis and one for resiliency analysis, considering smart grid SCADA systems. These frameworks take smart grid configurations and organizational security or resiliency requirements as inputs, formally model configurations and various security properties, and verify the dependability of the system under potential attacks or contingencies. The execution of each of these frameworks is demonstrated on an example case study.

Introduction

Many cyber-physical systems (CPSs) like smart power grids, transportation systems, and water treatment plants are identified as critical infrastructures (CIs) due to their national importance. Secure and dependable operations of these infrastructures are extremely important. One or more industrial control systems (ICSs) are often found in a CPS, which are responsible for optimally and efficiently managing the system in

M. A. Rahman (✉)
Florida International University, Miami, FL, USA
e-mail: marahman@fiu.edu

E. Al-Shaer (✉)
University of North Carolina at Charlotte, Charlotte, NC, USA
e-mail: ealshaer@uncc.edu

© Springer Nature Switzerland AG 2019
C. Rieger et al. (eds.), *Industrial Control Systems Security and Resiliency*, Advances in Information Security 75, https://doi.org/10.1007/978-3-030-18214-4_4

real-time. The supervisory control and data acquisition (SCADA) system is the most important kind of ICS, which is responsible for monitoring and controlling dispersed assets by gathering and analyzing real-time data from remote (field) devices. Typical ICS operations include automated control loops, human machine interfaces (HMIs), and remote diagnostics and maintenance utilities.

In order to promote connectivity and remote access capabilities among corporate business systems, information technology (IT) is now increasingly used in ICSs, which escalates the possibility of cyber security vulnerabilities and incidents. Although there are some similarities between the characteristics of ICS and that of traditional IT systems, they differ in many places, especially due to the simultaneous existence of physical components and network components, along with different industrial communication protocols. That is why the vulnerabilities and threats, as well as the security requirements, of an ICS are often different from that of traditional IT systems. As such, it is important to develop automated security and resiliency analytics specifically for ICSs.

In this chapter, two formal frameworks – one for security analysis and the other for resiliency analysis – are presented providing a theoretical base for assessing the security and resiliency of ICS. These frameworks automatically and provably analyze the security and resiliency of the SCADA system, particularly, in terms of the data acquisition for executing control operations in smart grids. The frameworks take necessary SCADA configurations and security/resiliency requirements, formally model the analytics, and solve the models to verify the system with respect to the given security or resiliency specifications. The formal models are solved using state-of-the-art logical solvers. Each framework provides threat vectors, specifying when the security (or resiliency) requirements fail under the attack model. The unsatisfiable outcome can certify that the system is secure (or resilient) against the attack model. These frameworks can allow a grid operator to understand a SCADA system's resiliency, as well as to fix the system, by analyzing the threat vectors.

State of the Art of Research, Challenges, and Solutions

With the rise of cyber-warfare, the secure and dependable operation of a smart grid carries utmost importance. However, due to its many cyber and physical couplings, realizing the extent of security and resiliency of the system is challenging.

Supervisory Control and Data Acquisition Systems

An example topology of a SCADA system is shown in Fig. 1. Typical SCADA operations include automatic and human control loops, remote diagnostics, and maintenance utilities. There are also various kinds of physical devices, such as SCADA control servers or master terminal units (MTUs), remote terminal units (RTUs), programmable logic controllers (PLCs), intelligent electronic devices

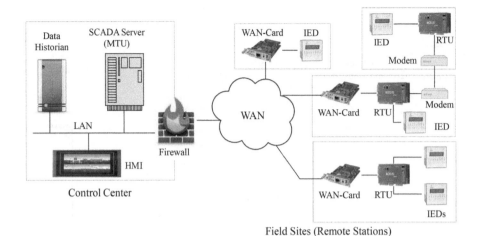

Fig. 1 An example of the SCADA network topology

(IEDs), human machine interfaces (HMIs), data historians, etc. IEDs, RTUs, and PLCs are considered as field or end devices, while the other devices reside in the control center. In addition to these control components, there are different network components, such as communications routers, modems, and remote access points. These components usually use ICS protocols like Modbus, DNP3, or IEC 61850 variants for communicating with one another.

The SCADA control server/MTU takes the sensor measurements from field devices through the power network and sends the control commands to them after analyzing the data using the same infrastructure. Control decisions are optimally made by the energy management system (EMS) by running several interdependent control modules or routines, namely, state estimation (SE), topology processor (TP), optimal power flow (OPF), contingency analysis (CA), and automatic generation control (AGC) [1, 2]. Executions of these EMS control routines are actively dependent on the data acquisition from the field devices. Among these modules, state estimation is the core component. Its function is to compute the unknown state variables of the power system from the sensor measurements received through the SCADA system. The output of state estimation is used in other control mechanisms to operate the grid optimally with respect to the generation cost and the physical safety of the grid. Figure 2 presents the core EMS modules and the interdependency among them according to the data flow.

Potential Cyber-Threats on SCADA

The increasing use of IT in smart grids escalates the possibility of cyber security vulnerabilities and incidents, as these systems have not been built taking security into consideration in the first place. The inherent complexity associated with

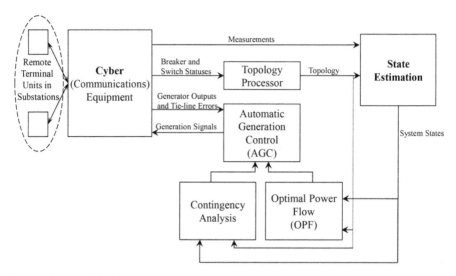

Fig. 2 A simplified EMS architecture. (Adapted from Ref. [1])

integrating different heterogeneous and legacy systems in SCADA systems signif-icantly increases the potential of security threats, which can cause massive and devastating damage. There are two main causes of threats [3]. The first is the *misconfiguration or the lack of security controls* that can cause inconsistency, unreachability, broken security tunnels, and many other security breaches. The second is the *weakness or absence of resiliency controls* that can lead to cascaded failures in contingencies or cyber-attacks. As an example of cyber-attacks, denial of service (DoS) attacks can make one or more field devices unreachable or unavailable to or from the rest of the system.

The main purpose of a SCADA system is to deliver measurement data from the field or physical devices (meters/sensors) to the provider's side (control center or utility) while delivering control commands from the provider's side to the field/ physical devices. To achieve successful data delivery, reachability must hold between the sender and the receiver. Inconsistencies in communication protocols or authentication/encryption parameters of the communicating devices may cause failed data transmission leading to service disruptions. In addition, data should be delivered such that it satisfies end-to-end integrity. The violation of this requirement not only can cause incorrect estimation of the system but may also launch malicious control commands toward physical devices. This scenario becomes worse in the case of contingencies, when some IEDs or RTUs fail due to technical errors or cyber-attacks, as there may not be enough (secured) measurements received by the control server to observe the whole system accurately.

Research Challenges and Formal Frameworks

The correct functioning of a SCADA system stands on consistent and secure execution of tasks in time. The safe security configuration depends not only on the local device parameters but also on the secure interactions and flows of these parameters across the network including SCADA control mechanisms. There are a significant number of logical constraints on configuration parameters of many SCADA devices, which need to be satisfied to ensure safe and secure communications among SCADA components while keeping the system stable during contingencies. The adversary must be modeled with respect to practical properties so that a realistic picture of the system's resiliency can be realized. The attack model needs to be flexible enough to consider a wide range of different attack scenarios. Implementing these security and resiliency controls in a scalable and provable manner is one of the major challenges in smart grid security modeling.

To address this grand challenge, formal frameworks are proposed that can allow energy providers to objectively assess and investigate SCADA security configurations to identify potential resiliency threats and to enforce smart grid operational and organizational security requirements [4–6]. These works primarily model secured communication, potential contingencies, and security/resiliency properties and provide an efficient solution to analyze the security and resiliency of the system by identifying the threat vectors that negate the security and resiliency requirements. The frameworks are designed as a constraint satisfaction problem. Although these frameworks include the formalizations of a limited set of constraints that are important for proper communication, an important feature of these frameworks is their easy extensibility. For further properties, one just needs to add formalizations for corresponding constraints.

Threat Analysis Architecture

The basic architecture for the formal threat analytics is shown in Fig. 3. The threat analyzer takes different inputs, including the bus/SCADA topology information (including connectivity between buses/SCADA devices), device configuration (including encryption and authentication properties, recorded measurements, etc.), and the control functions and corresponding data requirements. The analyzer also takes an attack model, along with a set of adversary attributes, as input. While the attack model specifies different kinds of attacks (e.g., false data injection attacks on measurements as well as topology statuses), the adversary attributes include, but are not limited to, (i) the attacker's knowledge of the system (e.g., the measurements location, topology, etc.); (ii) the attacker's capabilities for accessing and manipulating/compromising specific cyber/physical entities for launching attacks; and (iii) the attacker's resources, such as the potential for corrupting different physical entities at time-to-launch coordinated attacks. Based on this input, the security analyzer derives

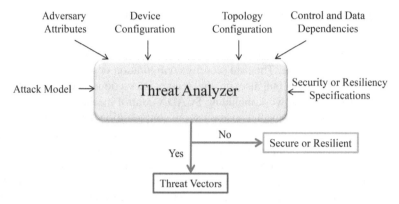

Fig. 3 The architecture of the threat analytics

the invariant properties of the physical system model, the attack model and potential evasion, and the inter-module dependencies and verifies feasible attacks/threats and provides corresponding threat vector(s) (i.e., states/measurements to be compromised or devices to be failed). The threat analysis modeling considers the interaction among the different EMS modules (i.e., SE, OPF, TP, CA, and AGC) in a way that can verify where and how an attack can be launched (e.g., SE and TP) and how far the attack (i.e., its impact) can percolate.

Formal Approach Characteristics

The formal security and resiliency frameworks [4–6] embrace the following key ideas, in general:

Security-Centric Modeling A security-centric model of the power system is one that explicitly integrates security properties into the core physical model. The model allows measurements/statuses received/sent from/to field devices through cyber-communication to be intentionally corrupted by adversaries. The model incorporates security properties into the entire measurement set by defining a set of variables to denote the cyber-physical and adversarial properties of the system.

Comprehensive Modeling of Adversary Attributes The adversary or attack attributes are expressed in terms of the knowledge about the target system, accessibility to the system, attack objectives, the resources to launch an attack, etc. The modeling needs to construct a formal description to map the changes that occurred on physical properties with respect to attacks. Modeling interdependencies among EMS modules will allow mapping of how the effect of an attack on a module can percolate to another module.

Unified Framework to Identify Coordinated Attack This step unifies the physical properties, attack model, and adversarial components into a comprehensive model for the entire complex system. The resulting unified formal model, expressed as a set of constraint satisfaction verification problems, is solved for the potential attacks. Interdependency modeling helps identify coordinated attacks (e.g., coordinating false data injections on measurements, as well as topology statuses).

Efficient Solutions by Satisfiability Modulo Theories (SMT) Proposed models are formalized using SMT. Over the past 10 years, SMT has become the core engine behind many practical tools for software and hardware analytics for static software analysis, dynamic symbolic execution, model-based testing, and automated synthesis and planning [7]. SMT is a constraint satisfaction problem solver for logical formulas with respect to combinations of background theories (e.g., uninterpreted functions, linear and non-linear arithmetic, difference logic, etc.). The SMT formula can be considered as an instance of the Boolean satisfiability problem (SAT) in which some of the binary variables are replaced by binary-valued predicates over a set of non-binary variables. SMT solvers can determine the satisfiability of formulas that contain thousands of variables and constraints [7].

Formal Framework for SCADA Security Analysis

The security analysis framework focuses on formally modeling the EMS control routines, their interdependency, and how false data injection can alter the outcome of the control routines – in particular SE and TP – without being detected by the traditional bad data detection algorithms. The mathematical formulation of stealthy attacks against SE is introduced by Liu et al. in 2009 [8], which has received the attention of many researchers since then.

Methodology

The steady-state physical properties of the grid are governed by power flow equations, which express the conservation relations between generation and load at every bus or node in the system, at every instant. In this project, EMS modules are formally modeled, particularly to determine the means (attack vectors) of stealthy attacks. In the following, a simplified model of the security analyzer, which is based on the linear power flow equations (the DC model), is presented. The overall formal model capturing the physical system, the cyber-physical attack properties, and adversary attributes are summarized in Table 1. A brief explanation of the salient steps leading to the formalization is provided below.

Table 1 DC power flow equations

#1: Physical power flow properties:	
Power flows and topology:	
$\forall_{1\le i\le l}\quad P_i^L = d_i\left(\theta_{f_i} - \theta_{e_i}\right)$	(1)
$\forall_{1\le i\le l}\quad k_i \rightarrow \left(P_i^L = d_i\left(\theta_{f_i} - \theta_{e_i}\right)\right)$	(2)
Power consumptions:	
$\forall_{1\le j\le b}\quad P_j^B = \sum_{i\in L_{j,in}} P_i^L - \sum_{i\in L_{j,out}} P_i^L$	(3)
Power generation, loads, and consumption relationship:	
$\forall_{1\le j\le b}\quad P_j^B = P_j^D - P_j^G$	(4)
$\sum_{1\le j\le b} P_j^G = \sum_{1\le j\le b} P_j^D$	(5)
#2: False data injection attack properties:	
Attack definitions:	
$\forall_{1\le j\le n}\quad c_j \rightarrow (\Delta\theta_j \ne 0)$	(6)
$\forall_{1\le i\le l}\quad k_i \rightarrow (u_i \wedge \neg p_i) \vee (\neg u_i \wedge q_i)$	(7)
$\forall_{1\le i\le l}\quad (p_i \rightarrow u_i \wedge \neg v_i \wedge \neg w_i) \wedge (q_i \rightarrow \neg u_i \wedge \neg w_i)$	(8)
Attack evasion (UFDI) properties:	
$\forall_{1\le i\le l}\quad \neg k_i \rightarrow \left(\Delta P_i^L = 0\right)$	(9)
$\forall_{1\le i\le l}\quad \neg(p_i \vee q_i) \rightarrow \left(\Delta\bar{P}_i^L = 0\right)$	(10)
$\forall_{1\le i\le l} k_i \rightarrow \left(\Delta P_i^L = d_i\left(\Delta\theta_{f_i} - \Delta\theta_{e_i}\right)\right)$	(11)
$\forall_{1\le i\le l}\quad \left(p_i \rightarrow \left(\Delta\bar{P}_i^L = -P_i^L\right)\right) \wedge \left(q_i \rightarrow \left(\Delta\bar{P}_i^L = P_i^L\right)\right)$	(12)
Attack plan properties:	
$\forall_{1\le i\le l}\quad \Delta P_{i,\text{total}}^L = \Delta P_i^L + \Delta\bar{P}_i^L$	(13)
$\forall_{1\le j\le b}\quad \Delta P_{j,\text{total}}^B = \sum_{i\in L_{j,in}} \Delta P_{i,\text{total}}^L - \sum_{i\in L_{j,out}} \Delta P_{i,\text{total}}^L$	(14)
$\forall_{1\le i\le l}\quad \left(\Delta P_{i,\text{total}}^L \ne 0\right) \rightarrow (t_i \rightarrow a_i) \wedge (t_{l+i} \rightarrow a_{l+i})$ $\forall_{1\le j\le b}\quad \left(\Delta P_{j,\text{total}}^B \ne 0\right) \rightarrow \left(t_{2l+j} \rightarrow a_{2l+j}\right)$	(15)
#3: Adversary attribute properties:	
Attacker's knowledge:	
$\forall_{1\le i\le l}\quad \left(\Delta P_{i,\text{total}}^L \ne 0\right) \rightarrow ((t_i \vee t_{l+i}) \rightarrow g_i)$	(16)
Attacker's access capability:	
$\forall_{1\le i\le m}\quad a_i \rightarrow r_i \wedge \neg s_i$	(17)
Attacker's resource:	
$\sum_{1\le i\le m} a_i \le T_M$	(18)

Physical Model

Power Flow Model The DC power flow model makes several simplifying assumptions [1] that yield a linear relation between a system of equations of the form: **[B]** $[\theta] = $ **[P]**. Here, P denotes a vector of new power injections at a bus (node), while θ denotes the phase angles of unit magnitude bus voltages. The later variables are treated as *states*. Equations 1, 2, 3, 4, and 5 represent the modeling of the power flow properties, where b and l are the number of buses and lines, respectively. Denoting the admittance (reciprocal of the impedance) of line i between buses f_i and e_i by d_i the real power flow $\left(P_i^L\right)$ across the line is represented by Eq. 1. The topology status (processed by TP based on the information of the circuit breakers and switches) is considered using k_i to denote if the line is open or closed (Eq. 2). The power consumption at bus j $\left(P_j^B\right)$ is the summation of the power injections at the bus (Eq. 3, where $L_{j,\text{in}}$ and $L_{j,\text{out}}$ are the sets of incoming lines and outgoing lines of bus j, respectively).

This consumption is also the difference of the load $\left(P_j^D\right)$ and the generation $\left(P_j^G\right)$ at the bus (Eq. 4). As the DC flow model assumes a lossless system, the power balance constraint (total generation = total load) is given by Eq. 5.

State Estimation The state estimation is the process of estimating n unknown variables (states) from m $(m > n)$ known measurements $\mathbf{z}$ assuming a system of the form [2, 9]:

$$\mathbf{z} = \mathbf{h}(\mathbf{x}) + \mathbf{e}$$

With the DC power flow model, this reduces to a linear model of the form: $\mathbf{z} = \mathbf{H}(\mathbf{x}) + \mathbf{e}$ where $\mathbf{H}$ is a $(m \times n)$ constant matrix. In simple words, this estimation is the process of solving line power flow and consumption equations (i.e., Eqs. 1 and 3) based on the received measurements for P_i^Ls and P_j^Bs for θ_js, based on the estimated topology (Eq. 2).

Cyber-Physical Attack Model

Idea of Stealthy Attacks Measurements can be corrupted due to device errors or communication noise, while there are bad data detection (BDD) algorithms to filter them. The widely used weighted least squares (WLS)-based BDD algorithm identifies a measurement as bad if the difference between a received measurement and its corresponding estimation (calculated from the estimated states) is greater than a threshold value. However, Liu et al. have shown that it is possible to compromise the state estimation by injecting false data to measurements while evading the BDD algorithm [8]. This type of stealthy attack is based on the idea of altering measurements following the physical properties (i.e., Eqs. 1 and 3). Table 1 presents the constraints that define attacks on measurements and topology (Eqs. 6, 7, and 8), the evasion or stealthy properties (Eqs. 9, 10, 11, and 12), and the attack plan (Eqs. 13, 14, and 15). This attack modeling is based on the difference between the original (attack free) and the corrupted (under attack) measurement values, which makes it possible to verify the attacks without knowing the actual measurements. Since state estimation is done based on the estimated topology, an adversary can poison the topology, along with measurements, by injecting a false status, leading to exclusion/inclusion of one or more open/closed lines from/to the topology.

Attack Modeling An attack on state j (c_j) specifies that θ at bus j is changed ($\Delta\theta_j$) (Eq. 6). In an "exclusion" attack (p_i), a line actually in service (u_i) is omitted, if the line is not a fixed one (v_i) and the status information corresponding to the line is not secured (w_i) (Eq. 8). An "inclusion" attack includes a line that is actually not in service, and its associated status information is not secured (Eq. 8). A line is considered (k_i) in the EMS routines if (i) the line is in service and no exclusion attack is launched against this line or (ii) an inclusion attack is performed on an open line (Eq. 7).

Attack Evasion Properties In order to evade the BDD algorithm, the injection of false data in the measurements $\left(\Delta P_i^L\right)$ should follow the change in states along with the rest of the measurements and states (Eq. 11). In the topology poisoning attack alone, the BDD algorithm is evaded by keeping the states unchanged, while necessary measurements are changed accordingly $\left(\Delta\bar{P}_i^L\right)$. If a closed line is excluded from the topology, the corresponding line power flow measurement must be zero and the corresponding connected buses' power consumption measurements are adjusted accordingly (Eq. 12). On the other hand, when an open line is included in the topology, there should be a non-zero line power flow according to the phase difference between the connected buses (Eq. 12).

Attack Vectors The attack vector (plan) is a set of measurements that need to be altered to coordinate the attack actions for stealthiness. The total change to a measurement is denoted by $\Delta P_{i,\text{total}}^L$ and $\Delta P_{j,\text{total}}^B$ (Eqs. 13 and 14). When $\Delta P_{i,\text{total}}^L \neq 0$, taken measurements corresponding to line i (i.e., t_i and t_{l+i}) are required to be altered (a_i and a_{1+i}) (Eq. 15) according to $\Delta P_{i,\text{total}}^L$. Similarly, when $\Delta P_{j,\text{total}}^B \neq 0$, the power consumption measurement at bus j needs to be changed (Eq. 15). Conversely, a measurement is altered only if it is required. A violation of these constraints will make the attack detectable.

Modeling Adversary Attributes

An adversary cannot have an unlimited capability, or a system cannot practically be secured from all possible attacks. Therefore, it is prudent to analyze an adversary with limited but practical capabilities (e.g., expressed in terms of knowledge, access, and resources). These capabilities can be expressed as formal constraints. With regard to knowledge, since the electrical characteristics of the grid and other system properties are usually well-guarded, they are not easily accessible to adversaries. If the admittance of a line is unknown (g_i), then an adversary cannot determine appropriate changes to power flow measurements (Eq. 16). The attacker usually does not have necessary physical or remote access (r_i) to inject false data into all of the measurements. If a measurement is secured (i.e., integrity-protected) (s_i), then even if the attacker has the ability to inject false data into the measurement, the false data will be detected (Eq. 17). Due to resource constraints, an adversary can corrupt a limited number of measurements/buses at a time (Eq. 18).

Interdependency Models

An example is presented here that formally models interdependency toward assessing the impact on an EMS module, namely, OPF, through attacking SE. OPF is responsible for determining individual generator outputs that minimize the overall cost of generation while meeting physical properties (i.e., transmission, generation, and system-level operating constraints) [2, 9]. Since an attack on state estimation can result in a redistribution of loads, an OPF solution may be no longer optimal, leading to an economically disadvantageous solution. The OPF considers the entire set of power flow equations as one big constraint, which includes the constraints regarding load-generation balance, generation limits, and transmission line capacities, along with cost minimization. Thus, OPF can be expressed as a conjunction of a set of individual constraints, while it can be merged with the attack models, with respect to the load changes.

Change in Loads Due to Stealthy Attacks If $\Delta P_j^B \neq 0$ (Eq. 4), this specifies that there is a load and/or generation power change at the bus. Since the generation (metered at the plant) is typically altered at the request of the system operator, it can be assumed that the power consumption change specifies a change exclusively in the load:

$$\forall_{1 \leq j \leq b} \ \Delta P_{j, \text{total}}^D = \Delta P_{j, \text{total}}^B$$

If a load change is observed, the OPF process must be rerun to find the optimal generation dispatches.

Impact on OPF The minimization function of OPF can be abstracted as a cost constraint:

$$\sum_{1 \leq j \leq b} c_j(\hat{P}_j^G) \leq T_{\text{OPF}}$$

Here, C_j is the cost function for the generation at bus j and T_{OPF} is the OPF cost threshold. Assuming that the cost (T_{OPF}) is increased with respect to the original (no-attack scenario) OPF cost T_{OPF} by $I\%$, the minimum impact on OPF can be expressed as:

$$(T_{\text{OPF}} = T_{\text{OPF}} \ I/100) \rightarrow \neg(\exists_{\hat{P}_1^G, \hat{P}_2^G, ..., \hat{P}_b^G} \text{OPF})$$

This constraint specifies that no OPF solution is possible that does not increase the generation cost by $I\%$.

Example Case Study

A case study of security threat analysis is briefly presented here [4, 5]. The system configuration and the constraints corresponding to the prior model are encoded into SMT [7] using the Z3 [10] solver. Boolean variables are used for logical constraints, while real variables are used for values of measurements and states. By executing the model, if the result is unsatisfied (*unsat*), then no attack vector can satisfy the constraints. However, if result is satisfied (*sat*), the attack vector is received from the assignments of the corresponding variables, which represent the measurements required to be altered to achieve the attack.

Attack Verification [4] This case study is to demonstrate stealthy attack verification on the IEEE 14-bus test system [11], as shown in Fig. 4. In this example, the admittances of lines 3, 7, and 17 are unknown. All of the 20 lines (as shown in Fig. 4) are included in the true topology, though lines 5 and 13 are not part of the core topology (i.e., these lines can be kept open if necessary). Since this system has 14 buses and 20 lines, the maximum number of potential measurements is (14 + 2 × 20) or 54. Here, all the potential measurements are used except measurements 5, 10, 14, 19, 22, 27, 30, 35, 43, and 52, and among these measurements, 1, 2, 6, 15, 25, 32, and 41 are crypto-secured for integrity protection. In this example, let's assume that the attacker has access to all measurements and the target is to attack state 12 only (i.e., no other states will be affected). However, due to the resource

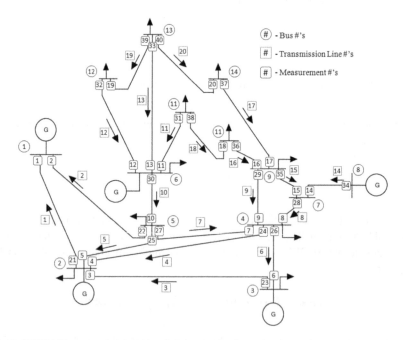

Fig. 4 IEEE 14-bus test system with measurement numbers

limitation, the attacker can alter up to 10 measurements distributed in three or less substations (i.e., buses). Is there a feasible attack vector in this case? The preliminarily implemented model (security analyzer) shows that attack for state 12 is feasible under this model when, for example, measurements 12, 32, 39, 46, and 53 are compromised (alerted). However, if only measurement 46 is considered as crypto-secured, then no feasible attack exists unless the attacker can alter (poison) the topology information as well. Specifically, the model can also tell if line 13 needs to be deceivably excluded from the topology by corrupting the topology information. This attack is feasible by compromising measurements 12, 13, 32, 33, 39, and 53. This is an example of novel attacks showing the capability of the proposed framework that allows the discovery of coordinated attacks (e.g., in this case, by integrating the topology poisoning with false data injection to measurements).

Impact-Based Attack Verification [5] The attack vector can be conditioned with the target impact. For example, if the adversary's objective is to launch a stealthy (undetected) attack to induce at least a 5% increase to the OPF base solution on the same 14-bus test system and the adversary can attack 25 measurements distributed on 10 buses, then the model identifies a feasible attack vector at which (1) states 10, 11, 12, and 14 are compromised through injecting false data to measurements 4, 11, 12, 16, 17, 18, 20, 24, 31, 32, 36, 37, 38, 39, 40, 42, 44, 46, 49, 50, 51, 53, and 54 and (2) the topology information is corrupted to falsely remove lines 4 and 17. The model can provide valuable insight on the required attacker profile and capabilities to accomplish an attack. For example, the model can confirm that, under any circumstance, an attacker with the same capability cannot cause 7% or more upward drift in OPF outcome even with coordinated false data injection (stealthy) attacks on the measurements and the topology.

Formal Model for SCADA Resiliency Analysis

The resiliency requirements that are considered in the resiliency threat analyzer [6] ensure whether or not a SCADA control process receives sufficient data (i.e., measurements from field devices) to perform its operation even in (limited) contingencies. The threat analyzer solves a formal model that formalizes necessary control function data requirements and security properties, along with the physical topology and devices.

Methodology

The observability analysis is a prior and crucial requirement for performing the power system state estimation control routine [1, 2]. The threat analysis is performed considering this observability analysis. Three resiliency specifications are modeled in this analyzer: (i) k —resilient observability; (ii) k—resilient secured observability; and (iii) k, r—resilient bad data detectability. A brief description of the formal model for k—resilient secured observability analysis is presented below.

SCADA Cyber-Physical System Modeling

The SCADA system modeling primarily includes the configurations associated with various SCADA devices and the topology, as follows:

SCADA Device Configuration Modeling A SCADA system consists of different physical device components, among which IEDs, PLCs, RTUs, and MTUs are important. The SCADA physical devices are modeled mainly based on their communication and security configurations, especially those properties that are essential to model the security specifications and resiliency requirements. Each SCADA device (e.g., an IED) is identified by an ID (e.g., i). Whether a device with an ID is an IED is determined using a parameter (e.g., Ied_i). Similar parameters are there for other SCADA devices. A parameter (e.g., Ied_i) can define whether device i is an IED. A device profile is represented as a conjunction of different parameters. The data reporting often follows a schedule and it is modeled based on the reporting mode. The reporting mode for the field devices can be pull or push, although a SCADA device typically delivers data to the control server upon receiving a request from the server. To achieve end-to-end security, the communicating devices must be configured with necessary cryptographic (authentication and encryption) properties. However, a device can support none, one, or multiple crypto properties. The crypto property of a device (e.g., $Crypt_i$ for device i) can be modeled as a conjunction of one or more crypto profiles ($CryptType_{i,k}$ for one or more ks). For example, each crypto profile (K, e.g., $CryptType_{i,k} = K$) specifies an algorithm ($CAlgo_K$) and a key length ($CKey_K$). Similarly, the communication properties (e.g., supported communication protocols) for each device are modeled.

SCADA Topology Modeling This task models the SCADA communication network topology (i.e., the connectivity between the SCADA physical devices). Typically, multiple IEDs or PLCs are connected with an RTU, while all or some RTUs are connected to an MTU directly or through some intermediate RTUs using WAN. However, different topology patterns are possible. There can be more than a single MTU, in which case one of them works as the main MTU (corresponding to the main control center), while the rest of the MTUs are hierarchically connected to the main one. The measurements and control commands route through this communication topology between the devices. Although the communication among field devices in SCADA often can be point-to-point (e.g., an IED to an RTU or an RTU to an RTU), this modeling will consider intermediate network devices like routers and firewalls when they exist. A link in the topology is identified by an ID (e.g., l), while a parameter (e.g., $NodePair_l$) represents the nodes connected by the link, and a parameter (e.g., $LinkStatus_l$) specifies if the link is up or down. There can be other properties, including the medium type (i.e., wireless, Ethernet, modem, etc.) and the link bandwidth. It is worth mentioning that a communication path (e.g., a routing path through routers and links from an RTU to another RTU) can be abstracted as a link as long as the internal routing path is not considered for a resiliency specification.

Modeling of Attacks and Security Controls

The attack model is designed with respect to SCADA networks. The design of security controls primarily deals with end-to-end data communication.

Attack Model Design Cyber-attacks corresponding to data unavailability are modeled here. A particular data can be unavailable due to some technical failures at the source node, intermediate forwarding nodes, or communication links, as well as due to distributed denial of service (DDoS) attacks. A Boolean parameter ($Node_i$) is used to denote whether device i is available or not.

Security Control Modeling A requirement analysis is performed on standard security recommendations for SCADA (e.g., NIST and NERC security guidelines [12, 13]) to identify logical structures associated with SCADA configurations and security properties. The major control modeled here is secured data delivery.

The secured data delivery checks for assured data delivery, as well as whether the data is sent under proper security measures, particularly authentication and integrity protection. The data delivery is ensured with the satisfaction of various constraints, which primarily include these three: (i) reachability, (ii) communication protocol pairing, and (iii) crypto property pairing. The communicating nodes (e.g., an RTU and the MTU) may have correct security pairing, as they are using the same security protocol (e.g., challenge-handshake authentication protocol (CHAP)). However, this security pairing on CHAP can only ensure authentication. In this case, the transmission will not be data integrity-protected. Moreover, it is needed to consider the vulnerabilities of the security measures in use. For example, if data encryption standard (DES) is used for data encryption, the transmitted data cannot be considered as protected, as a good number of vulnerabilities of DES have already been found. Hence, the formalization of the secured data delivery ($SecuredDelivery$) includes two constraints – $Authenticated$ and $IntegrityProtected$ – that ensure the authentication of the communicating parties and the integrity of the transmitted data, respectively:

$$\exists_K \left(\exists_K CryptType_{i,k} = K \right) \wedge \left(\exists'_k CryptType_{i,k'} = K \right) \wedge$$

$$\left(\left(CAlgo_K = hmac \wedge CKey_K \geq 128 \right) \vee \ldots \right) \rightarrow Authenticated_{i,j}$$

$$\exists_K \left(\exists_K CryptType_{i,k} = K \right) \wedge \left(\exists'_k CryptType_{i,k'} = K \right) \wedge$$

$$((CAlgo_K = sha2 \wedge CKey_K \geq 128) \vee \ldots)) \rightarrow IntegrityProtected_{i,j}$$

$$Ied_l \wedge \exists_z \forall_{l \in |P_{l,j,z}} \{i',j'\} \in NodePair_l \wedge Node_{i'} \wedge Node_{j'} \wedge$$

$$Reachable_{i',j'} \wedge CommPropPairing_{i',j'}, \wedge CryptoPropPairing_{i',j'} \wedge$$

$$Authenticated_{i',j'}, \wedge IntegrityProtected_{i',j'}$$

$$\rightarrow \quad SecuredDelivery_l$$

Modeling of Resiliency Threats Based on SCADA Operations

An essential resiliency requirement in general can ensure whether or not a SCADA control process receives sufficient (secured) data (i.e., measurements from field devices) to perform its operation, even if some contingency occurs within some limit (e.g., a threshold number of field devices are under failure due to data unavailability attacks). The resiliency threat verification is designed to answer the same query, in other words, by looking for a set of field devices (e.g., IEDs and RTUs) such that the set size is no more than a threshold value and the unavailability of these devices will make the SCADA control process fail because of insufficient data. The modeling of k—resilient secured observability property is performed as follows.

The property is verified by searching for threat vectors under the specification of maximum failures of k field devices. When the number of unavailable devices is no larger than k devices (IEDs and/or RTUs), the threat against the k—resilient secured observability requirement ($\neg ResilientSecuredObservability$) is formalized as follows:

$$\left(\left(N - \sum_{1 \leq i \leq N} Node_i \right) \leq k \right) \wedge \neg SecuredObservability$$

$$\rightarrow \quad \neg ResilientSecuredObservability$$

The resiliency requirement can specify the device type clearly (e.g., k_1 IEDs and k_2 RTUs), instead of k devices of any (or multiple) types. The threat against the (k_1, k_2)—resilient secured observability requirement is formalized as follows:

$$\left(\left(N_1 - \sum_{1 \leq i \leq N_1} (Node_i \times Ied_i) \right) \leq k_1 \right) \wedge$$

$$\left(\left(N_2 - \sum_{1 \leq i \leq N_2} (Node_i \times Rtu_i) \right) \leq k_1 \right) \wedge \neg SecuredObservability$$

$$\rightarrow \quad \neg ResilientSecuredObservability$$

The threat vector ($\mathcal{V}$) represents those devices for which the following equation is true: $\forall_{i \in \mathcal{V}} \neg Node_i$.

Example Case Study

The resiliency threat analyzer's execution is illustrated with an example. This example considers a 5-bus SCADA system as shown in Fig. 5 and demonstrates the k_1, k_2—resilient secured observability analysis. This is a subsystem taken from the IEEE 14-bus test system [11]. The input includes primarily the Jacobian matrix corresponding to the bus system [2], the connectivity between the communicating devices, the association of the measurements with the IEDs, and the security profiles of each communicating host pair. It is assumed that the measurements are recorded by different IEDs only, and these measurements are sent to the MTU (i.e., the SCADA server at the control center) through the RTUs. The server needs these measurements to estimate the current states of the system. The resiliency specification is 1, 1—resilient secured observability. The formal model corresponding to this

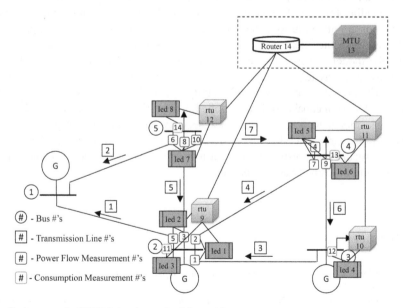

Fig. 5 An example SCADA topology of a 5-bus grid

example is solved using Z3 [10]. The solution to the model gives a result as *sat* (satisfiable) or *unsat* (unsatisfiable). In a *sat* case, the solver provides an elaborated result, specifically the values of the terms, from which the required output is assembled. In the case of (1, 1)—resilient secured observability verification, the model provides a *sat* result. That is, the system is not (1, 1)—resilient in terms of secured observability. According to the result, if IED 3 and RTU 11 are unavailable, it is not possible to observe the system securely. The result also justifies the answer, showing that measurements from IED 1 and RTU 9 are not data integrity-protected, and thus, when IED 3 and RTU 11 are unavailable, some states cannot be observed securely. By continuing to look for the next threat vector, another four threat vectors can be observed. However, if the resiliency specification is reduced to (1, 0) or (0, 1), the model returns *unsat* (i.e., the system is securely observable even if an IED or RTU fails). Now, if the SCADA topology of Fig. 5 is modified by connecting RTU 9 with RTU 10, while removing the direct path between RTU 9 and the MTU, the system is not resilient any more for one RTU failure. There is only one threat vector (unavailability of RTU 12) that fails the secured observability.

Conclusion

Unlike the existing approaches that focus on discovering specific attack vectors, the presented formal approach offers a comprehensive analysis for verifying SCADA security and resiliency properties systematically, provably, and efficiently. The corresponding frameworks take the bus data, SCADA device configurations, operational constraints, security properties, and resiliency requirements as inputs; formally model secure interactions among the devices and potential contingencies; and solve the model to verify the resiliency of the system. The key features of these frameworks are as follows: (i) a formal framework that utilizes advanced formal logics; (ii) provable verification of security and resiliency threats; (iii) a generic framework design capable of being applied in any SCADA architecture; and (iv) an extensible model to accommodate new security and resiliency properties. These formal frameworks can allow the grid operators to provably query and inspect the system's security and resiliency without relying on invasive, laborious, and expensive real-life or testbed-based experiments.

References

1. A.J. Wood, B.F. Wollenberg, *Power Generation, Operation, and Control*, 2nd edn. (Wiley, New York, 1996)
2. A. Abur, A.G. Exposito, *Power System State Estimation: Theory and Implementation* (CRC Press, New York, 2004)

3. Nistir 7628: Guidelines for smart grid cyber security. (Smart Grid Interoperability Panel- Cyber Security Working Group, Aug 2010), http://www.nist.gov/smartgrid/upload/nistir-7628_total. pdf
4. M.A. Rahman, E. Al-Shaer, R. Kavasseri. Security threat analytics and countermeasure synthesis for state estimation in smart power grids. In *IEEE/IFIP International Conference on Dependable Systems and Networks (DSN)*, June 2014
5. M.A. Rahman, E. Al-Shaer, R. Kavasseri. Impact analysis of topology poisoning attacks on economic operation of the smart power grid. In *International Conference on Distributed Computing Systems (ICDCS)*, July 2014
6. M.A. Rahman, A.H.M. Jakaria, E. Al-Shaer. Formal analysis for dependable supervisory control and data acquisition in smart grids. In *IEEE/IFIP International Conference on Dependable Systems and Networks (DSN)*, June 2016
7. L. de Moura, N. Bjørner. Satisfiability modulo theories: An appetizer. In *Brazilian Symposium on Formal Methods*, 2009
8. Y. Liu, P. Ning, M. Reiter. False data injection attacks against state estimation in electric power grids. In *ACM Conference on Computer and Communications Security (CCS)*, pp. 21–32, Nov 2009
9. A. Monticelli, *State Estimation in Electric Power Systems: A Generalized Approach* (Kluwer Academic Publishers, Norwell, 1999)
10. Z3: Theorem prover. (Microsoft Research, 2013), http://research.microsoft.com/en-us/um/red mond/projects/z3/
11. Power systems test case archive. http://www.ee.washington.edu/research/pstca/
12. National Institute of Standards and Technology. U.S. Department of Commerce. http://www. nist.gov/, http://www.nist.gov/publication-portal
13. North American Electric Reliability Corporation. http://www.nerc.com, http://www.nerc.com/ pa/Stand/Pages/default.aspx

Toward Cyber-Resiliency Metrics for Action Recommendations Against Lateral Movement Attacks

Pin-Yu Chen, Sutanay Choudhury, Luke Rodriguez, Alfred O. Hero, and Indrajit Ray

Abstract Lateral movement attacks are a serious threat to enterprise security. In these attacks, an attacker compromises a trusted user account to get a foothold into the enterprise network and uses it to attack other trusted users, increasingly gaining higher and higher privileges. Such lateral attacks are very hard to model because of the unwitting role that users play in the attack and even harder to detect and prevent because of their low and slow nature. In this chapter, a theoretical framework is presented for modeling lateral movement attacks and for designing resilient cyber-systems against such attacks. The enterprise is modeled as a tripartite graph capturing the interactions between users, machines, and applications, and a set of procedures is proposed to harden the network by increasing the cost of lateral movement. Strong theoretical guarantees on system resilience are established and experimentally validated for large enterprise networks.

Introduction

Cybersecurity is one of the most critical problems of our time. Notwithstanding the enormous strides that researchers and practitioners have made in modeling, analyzing, and mitigating cyber-attacks, black hats find newer and newer methods for launching attacks requiring white hats to revisit the problem with a new perspective.

P.-Y. Chen
IBM Thomas J. Watson Research Center, Yorktown Heights, NY, USA
e-mail: Pin-Yu.Chen@ibm.com

S. Choudhury (✉) · L. Rodriguez
Pacific Northwest National Laboratory, Richland, WA, USA
e-mail: sutanay.choudhury@pnnl.gov; luke.rodriguez@pnnl.gov

A. O. Hero
EECS, University of Michigan, Ann Arbor, MI, USA
e-mail: hero@eecs.umich.edu

I. Ray
Department of Computer Science, Colorado State University, Ft. Collins, CO, USA
e-mail: Indrajit.Ray@colostate.edu

© Springer Nature Switzerland AG 2019
C. Rieger et al. (eds.), *Industrial Control Systems Security and Resiliency*, Advances in Information Security 75, https://doi.org/10.1007/978-3-030-18214-4_5

One of the major ways [1] that attackers launch an attack against an enterprise is by what is known as "lateral movement via privilege escalation." This attack cycle, shown in Fig. 1, begins with the compromise of a single-user account (not necessarily a privileged one) in the targeted organization typically via phishing email, spear-phishing, or other social engineering techniques. From this initial foothold and with time on his/her side, the attacker begins to explore the network, possibly compromising other user accounts until he/she gains access to a user account with administrative privileges to the coveted resource: files containing intellectual property, employee or customer databases, or credentials to manage the network itself. Typically, the attacker compromises multiple intermediate user accounts, each granting him increasing privileges. Skilled attackers frequently camouflage their lateral movements into the normal network traffic making these attacks particularly difficult to detect and insidious.

Such lateral attacks are particularly insidious because authorized users play the role of an unwitting accomplice. End users have often been recognized as the "weakest links" in cybersecurity [1]. They do not follow security advice and often take actions that compromise themselves, as well as others. While efforts to educate and train end users for cybersecurity are important steps, anecdotal evidence shows that they have not been as effective. Clearly, there is a need for designing large enterprises that are resilient against such lateral movement attacks. Our current work takes a step in this direction.

Resilient systems accept that not all attacks can be detected and prevented; nonetheless, the system should be able to continue operation even in the face of cyber-attacks and provide its core services or *missions* even if in a degraded manner [2]. To build such a resilient system, it is important to be proactive in understanding and reasoning about lateral movement in an enterprise network, its potential effects

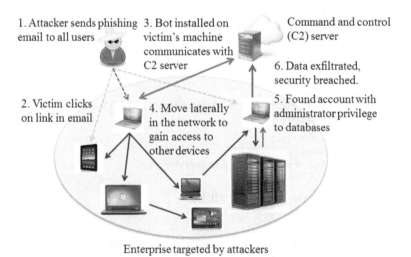

Fig. 1 An illustration of a cyber-attack using privilege escalation techniques

on the organization, and in identifying ways to best defend against these threats. Unfortunately, a theoretical framework for such risk analysis is currently missing. Our goal in this chapter is to establish the theoretical foundations of a systematic framework for building networks resilient to lateral movement attacks.

We model lateral movement attack on an enterprise's mission as a graph walk on a tripartite *user-host-application* network that logically comprises of two subgraphs: a *user-host* graph and a *host-application* graph. Figure 2 illustrates the model and our methodology. Specifically, Fig. 2a illustrates a tripartite network consisting of a set of users, a set of hosts, and a set of applications. Figure 2b illustrates the paradigm of resiliency via *segmentation*—the user, "Charlie," modifies his access configuration by disabling the access of his existing account (Charlie-2) to host H3, and by creating a new user account (Charlie-1) for accessing H3, such that an attacker cannot reach the data server H5 though the printer H3 if Charlie-2 is compromised. Figure 2c illustrates resiliency by network *edge-hardening* via additional firewall rules on all network flows to H5 through HTTP. Finally, Fig. 2d is an illustration of resiliency by *node-hardening* via system update or security patch installation on H5.

The user-host-application paradigm allows us to develop an abstraction of a mission in terms of concrete entities whose behavior can be monitored and

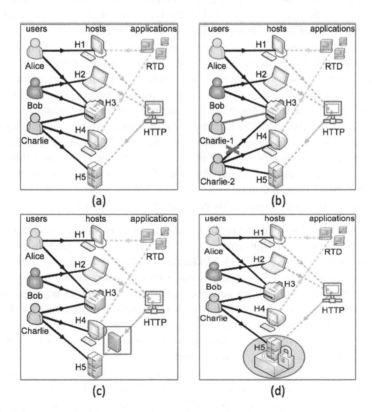

Fig. 2 Enterprise resiliency using user-host-application tripartite graph

Table 1 Utility of the proposed algorithms and established theoretical results

System heterogeneity	Hardening methods	Theoretical guarantees
User-host	Algorithms 1 and 2	Property 1, property 2, property 3
Host-application	Algorithms 3 and 4	Property 6, property 7
User-host-application	All of the above	All of the above

controlled, which captures interactions between diverse categories of users, software, and hardware resources (e.g., virtual machines, workstations, mobile devices) and applications.

Defining lateral movements as graph walks allows us to determine which nodes in the tripartite graph can be *reached* starting at a given node. From an attacker's perspective, the nodes that can be "reached" are exactly those mission components that can be attacked and compromised via exploits. The larger the number of nodes that can be reached by the attacker, the more "damage" he/she can cause to the mission. Given a system snapshot and a compromised workstation or mobile device, we can define the "attacker's reachability" as a measure that estimates the number of hosts at risk through a given number of system exploits. Now, from a defender's perspective, putting some defensive control on one of these nodes (or edges) allows the walk to be broken at that point. Intuitively, such a walk can also be used to identify mission-hardening strategies that reduce risk. This central idea is illustrated in Fig. 2. The heterogeneity of a cyber-system entails a network of networks (NoN) representation of entities in the system as displayed in Fig. 2, allowing us to devise effective hardening strategies from different perspectives, which differs from works focusing on manipulating the network topology under the assumption that the graph is homogeneous, that is, all nodes have an identical role in a cyber-system.

As our model considers the heterogeneity of a cyber-system and incorporates several defensive actions for enhancing the resilience to lateral movement attacks to assist reading the utility of the proposed approaches, the established theoretical results are summarized in Table 1. For lack of space, we omit the proofs of the established mathematical results.

The research contributions of this paper are listed as follows:

(a) By modeling lateral movements as graph walks on a user-host-application tripartite graph, we can specify the dominant factors affecting attacker's reachability (section "Network Model and Iterative Reachability Computation of Lateral Movement"), setting the stage for proposing greedy hardening and segmentation algorithms for network configuration change recommendation to reduce the attacker's reachability (sections "Segmentation on User-Host Graph" and "Hardening on Host-Application Graph").

(b) We characterize the effectiveness of three types of defensive actions against lateral movement attacks, each of which can be abstracted via a node or edge operation on the tripartite graph, which are (a) segmentation in the user-host graph (section "Segmentation on User-Host Graph"), (b) edge-hardening in host-application graph (section "Hardening on Host-Application Graph"), and (c) node-hardening in the host-application graph (section "Hardening on Host-Application Graph").

(c) We provide quantifiable guarantees (e.g., submodularity) on the performance loss of the proposed greedy algorithms relative to the optimal batch algorithms that have combinatorial computation complexity.

(d) We apply our algorithms to a collected real tripartite network dataset and demonstrate that the proposed approaches can significantly constrain an attacker's reachability and hence provide effective configuration recommendations to secure the system (section "Experimental Results").

(e) We collect traces of real lateral movement attacks in a cyber-system for performance evaluation (section "Performance Evaluation on Actual Lateral Movement Attacks"). We benchmark our approach against the NetMelt algorithm [1] and show that our approach can achieve the same reduction in attacker's reachability by hardening nearly 1/3 of the resources as recommended by NetMelt.

Background and Related Work

Laterally moving through a cyber-network looking to obtain access to an administrator's credentials or confidential information is a common technique in an attacker's toolbox [3]. Particularly, privilege escalation through lateral movement is a critical challenge for the security community [4–6]. For anomaly detection, Das et al. [7] employ graph clustering to group activities with a similar behavioral pattern and make change recommendations when the access control methods in place deviate from the real-world activity patterns. Chen et al. [8] use community structure to detect anomalous insiders in collaborative information systems. For attack prevention, Zheng et al. [9] use a graph partitioning approach to fragment the network to limit the possibilities of lateral movement. For risk assessment, Chen et al. [10] and Cheng et al. [11] use epidemic models for modeling and controlling malware propagation.

Our work fits into two emerging areas of study: (1) *network of networks* (NoN) representing multiple inter-related networks as a single model and (2) studies on the resilience of networks. Recently NoN has been an active area of research with diverse topics such as cascading failure analysis and control in interdependent networks [12, 13], improved grouping or ranking of entities in a network [14], and mapping of domain problems into the NoN paradigm [15]. Network resilience has been a long-studied topic. Demeester et al. [16] primarily focus on the physical topology of communication networks. There has been a surge in focus on enterprise-level cyber-resilience [2, 17], where the entire enterprise structure is modeled as a NoN.

Research has at times focused on altering the network structure to improve its resilience, as measured in terms of the spectral properties [18, 19]. Preventing contagion in networks is another attribute for resilience, and approaches such as that in Adamic et al. [20] suggest algorithms that *immunize* a subset of nodes as a preventive measure. Hu and Lau [21] contribute to this research area by unifying multiple data sources (e.g., different perspectives of user behaviors) into a single model. Integration of multiple data sources, such as user access control and application traffic, over the network makes the model more comprehensive and resulting recommendations more profound. The work is tailored toward providing action recommendations for enhancing the resilience of a heterogeneous cyber-system based on the associated NoN representation, which differs from previous work that focuses on manipulating the topology of a simple (homogeneous) network where each node in the graph has an identical role [1, 22]. To the best of our knowledge, the information in this chapter proposes the first representation of a cyber-system using the NoN model for designing algorithms that improve resiliency against lateral movement attacks.

Network Model and Iterative Reachability Computation of Lateral Movement

Notation and Tripartite Graph Model

Throughout this chapter, a scripted uppercase letter (e.g., $\mathcal{X}$) denotes a set, a boldface uppercase letter (e.g., $\mathbf{X}$ or $\mathbf{X}_k$) denotes a matrix, and its entry in the i-th row and the j-column is denoted by $[\cdot]_{ij}$; a boldfaced lowercase letter (e.g., $\mathbf{x}$ or $\mathbf{x}_k$) denotes a column vector, and its i-th entry is denoted by $[\cdot]_i$; and a plain uppercase or lowercase letter (e.g., X or x) denotes a scalar unless otherwise specified. The expression $|\mathcal{X}|$ denotes the number of elements in the set $\mathcal{X}$. The expression e denotes the Euler's number (i.e., the base of the natural logarithm). The expression $\mathbf{e}_i^x$ denotes the $x \times 1$ canonical vector of zero entries, except its i-th entry is 1. The expression $\mathbf{I}_n$ denotes the $n \times n$ identity matrix. The expression $\mathbf{1}_n$ denotes the n column vector of ones. The expression $\mathrm{col}_x(\mathbf{X})$ denotes the x-th column of $\mathbf{X}$. The expression $\lambda_{\max}(\mathbf{X})$ denotes the largest eigenvalue (in magnitude) of a square matrix $\mathbf{X}$. The operation $\cdot^{\mathrm{T}}$ denotes a matrix or vector transpose.

The operator $\otimes$ denotes the Kronecker product, which is defined as follows: Let $\mathbf{X}_1$ be a $r_1 \times l_1$ matrix and $\mathbf{X}_2$ be a $r_2 \times l_2$ matrix. The Kronecker product $\mathbf{X}_1 \otimes \mathbf{X}_2$ is a $r_1 r_2 \times l_1 l_2$ matrix defined as:

$$\mathbf{X}_1 \otimes \mathbf{X}_2 = \begin{bmatrix} [\mathbf{X}]_{11}\mathbf{X}_2 & [\mathbf{X}]_{12}\mathbf{X}_2 & \cdots & [\mathbf{X}]_{1l_1}\mathbf{X}_2 \\ [\mathbf{X}]_{21}\mathbf{X}_2 & [\mathbf{X}]_{22}\mathbf{X}_2 & \cdots & [\mathbf{X}]_{2l_1}\mathbf{X}_2 \\ \vdots & \vdots & \vdots & \vdots \\ [\mathbf{X}]_{r_11}\mathbf{X}_2 & [\mathbf{X}]_{r_12}\mathbf{X}_2 & \cdots & [\mathbf{X}]_{r_1l_1}\mathbf{X}_2 \end{bmatrix}$$

The operator $\odot$ denotes the Hadamard (entry-wise) product of matrices. The mapping $\mathbb{T} : \mathbb{R}_+^n \mapsto [0, 1]^n$ is a threshold function, such that $[\mathbb{T}(\mathbf{x})]_i = [\mathbf{x}]_i$ if $0 \le [\mathbf{x}]_i$

Table 2 Summary of main notations and symbols

U/N/K	Number of users/hosts/applications
$\lambda_{\max}(\mathbf{X})$	Largest eigenvalue of matrix $\mathbf{X}$
$\otimes$	Kronecker product
$\mathbf{1}_n$	$n \times 1$ column vector of ones
$\mathbf{A}_C$	User-host access matrix
$\mathbf{A}$	Host-application graph matrix
$\mathbf{P}$	Compromise probability matrix
$\mathbf{B}$	$\mathbf{A}_C^{\mathrm{T}}\mathbf{A}_C$
$\mathbf{J}$	$(\mathbf{P} \otimes \mathbf{1}_N)^{\mathrm{T}}\mathbf{A}^{\mathrm{T}}$
r/a	Reachability/hardening level vector
$\mathbb{T}(\mathbf{x})$	Threshold function on vector $\mathbf{x}$
$\mathbb{H}_a(\mathbf{x})$	Entrywise indicator function of $\mathbf{x}$ and $\mathbf{a}$

≤ 1 and $[\mathbb{T}(\mathbf{x})]_i = 1$ if $[\mathbf{x}]_i > 1$. The mapping $\mathbb{H}_a : [0, 1]^n \mapsto \{0, 1\}^n$ is an entry-wise indicator function, such that $[\mathbb{H}_a(\mathbf{x})]_i = 1$ if $[\mathbf{x}]_i > [\mathbf{a}]_i$ and $[\mathbb{H}_a(\mathbf{x})]_i = 0$ otherwise.

The tripartite graph in Fig. 2 can be characterized by a set of users, $\mathcal{V}_{\text{user}}$; a set of hosts, $\mathcal{V}_{\text{host}}$; a set of applications, $\mathcal{V}_{\text{app}}$; a set of user-host accesses, $\mathcal{E} \subset \mathcal{V}_{\text{user}} \times \mathcal{V}_{\text{host}}$; and a set of host-application-host activities, $\mathcal{T} \subset \mathcal{V}_{\text{host}} \times \mathcal{V}_{\text{app}} \times \mathcal{V}_{\text{host}}$. The cardinality of $\mathcal{V}_{\text{user}}$, $\mathcal{V}_{\text{host}}$, and $\mathcal{V}_{\text{app}}$ are denoted by U, N, and K, respectively. The main notation and symbols are summarized in Table 2.

Reachability of Lateral Movement on User-Host Graph

Let $G_C = (\mathcal{V}_{\text{user}}, \mathcal{V}_{\text{host}}, \mathcal{E})$ with $\mathcal{E} \subset \mathcal{V}_{\text{user}} \times \mathcal{V}_{\text{host}}$ denoting the user-host bipartite graph. The access privileges between users and hosts are represented by a binary $U \times N$ adjacency matrix, $\mathbf{A}_C$, where $[\mathbf{A}_C]_{ij} = 1$ if user i can access host j, and $[\mathbf{A}_C]_{ij} = 0$ otherwise. Let $\mathbf{r}_0$ be an $N \times 1$ binary vector indicating the initial host compromise status, where $[\mathbf{r}_0]_j = 1$ if host j is initially compromised and $[\mathbf{r}_0]_j = 0$ otherwise. Given $\mathbf{r}_0$, we are interested in computing the final binary host compromise vector $\mathbf{r}_\infty$ when attackers have leveraged user access privileges to compromise other accessible hosts. Note that lateral movement can be formally expressed a graph walk, that is, a sequence $v_0, e_1, v_1, \ldots, e_{k-1}, v_k$ of graph vertices v_i and graph edges e_i such that for $1 \leq i \leq k$ the edge e_i is incident on the vertices v_{i-1} and v_i. The vector $\mathbf{r}_\infty$ specifies the *reachability* of a lateral movement attack with reachability being defined as the fraction of hosts that can be reached via graph walks on G_C starting with the host compromise status expressed by $\mathbf{r}_0$. *Thus, reachability is used as a quantitative measure of enterprise network vulnerability to lateral movement attacks.* Moreover, studying $\mathbf{r}_\infty$ allows us to investigate the dominant factor that leads to high reachability and more efficient countermeasures.

The computation of $\mathbf{r}_\infty$ can be viewed as a cascading process of repetitive graph walks on G_C starting from a set of compromised hosts. Let $\mathbf{r}_t$ denote the binary compromise vector after t-hop walks, and let $\mathbf{w}_h$ be the number of h-hop walks starting from $\mathbf{r}_0$ and $\mathbf{w}_0 = \mathbf{r}_0$. The hop count of a walk between two hosts in G_C is defined as the number of traversed (affected) users. We begin by computing $\mathbf{r}_1$ from $\mathbf{r}_0$ as follows:

The number of 1-hop walk from $\mathbf{r}_0$ to host j is:

$$[\mathbf{w}_1]_j = \sum_{i=1}^{U}\sum_{k=1}^{N}[\mathbf{A}_C]_{ij}[\mathbf{A}_C]_{ik}[\mathbf{r}_0]_k$$
$$= \mathbf{e}_j^{N^T}\mathbf{A}_C^T\mathbf{A}_C\mathbf{r}_0$$

Let $\mathbf{B}=\mathbf{A}_C^T\mathbf{A}_C$ be an induced adjacency matrix of hosts in G_C where $[\mathbf{B}]_{ij}$ is the number of common users that can access hosts i and j. Then we have $\mathbf{w}_1 = \mathbf{B}\mathbf{r}_0$ and $\mathbf{r}_1 = \mathbb{T}(\mathbf{w}_1)$. Generalizing this result, we have:

$$\mathbf{w}_{h+1} = \mathbf{B}\mathbf{w}_h = \mathbf{B}^{h+1}\mathbf{r}_0 \tag{1}$$

$$\mathbf{r}_{t+1} = \mathbb{T}\left(\sum_{h=0}^{t+1}\mathbf{w}_h\right) \tag{2}$$

The term in Eq. (2) accounts for the accumulation of compromised hosts up to $t + 1$ hops. Note that based on the property of $\mathbb{T}$, Eq. (2) can be simplified as:

$$\mathbf{r}_{t+1} = \mathbb{T}(\mathbf{r}_t + \mathbf{B}\mathbf{r}_t) \tag{3}$$

The recursive relation of reachability in Eq. (3) suggests that the term, $\mathbf{B}$, is the dominant factor affecting the spread of lateral movement. Moreover from Eq. (3), we obtain an efficient iterative algorithm for computing $\mathbf{r}_\infty$ that involves successive matrix-vector multiplications until $\mathbf{r}_t$ converges.

Reachability of Lateral Movement on Host-Application Graph

The host-application graph contains information of one host communicating with another host via an application. Let $\mathbf{A}_k$ be an $N \times N$ binary matrix representing the host-to-host communication through application k, where $[\mathbf{A}_k]_{ij} = 1$ means host i communicates with host j by means of application k and $[\mathbf{A}_k]_{ij} = 0$ otherwise. The $N \times KN$ binary matrix $\mathbf{A} = \mathbf{A}_1\ \mathbf{A}_2\cdots\mathbf{A}_K$ is the concatenated matrix of K host-application-host matrices $\mathbf{A}_k$ for $k = 1, 2, \ldots, K$. Let $\mathbf{P}$ denote the compromise matrix, which is $K \times N$ matrix where its values $[\mathbf{P}]_{kj}$ specify the probability of compromising host j through application k. Each host is also assigned a hardening value $[\mathbf{a}]_j \in [0, 1]$ indicating the host's security level.

As in section "Reachability of Lateral Movement on User-Host Graph", we are interested in computing the spread of lateral movement on the host-application graph. The hop count of a walk between two hosts in the host-application graph is defined as the average number of paths between the two hosts via applications. Let $\mathbf{W}$ be an $N \times N$ matrix where $[\mathbf{W}]_{ij}$ is the average number of one-hop walk from host i to host j. Then we have $[\mathbf{W}]_{ij} = \sum_{k=1}^{K} [\mathbf{A}_k]_{ij} \mathbf{P}_{kj}$. Let $\mathbf{w}_h$ be an $N \times 1$ vector representing the average number of h-hop walks of hosts and $\mathbf{w}_0 = \mathbf{r}_0$. Then the jth entry of the 1-hop vector $\mathbf{w}_1$ is:

$$[\mathbf{w}_1]_j = \mathbf{e}_j^{\mathrm{T}} \left[\mathrm{col}_j(\mathbf{P})^{\mathrm{T}} \otimes \mathbf{I}_n \right] \mathbf{A}^{\mathrm{T}} \mathbf{r}_0 \tag{4}$$

Stacking Eq. (4) as a column vector gives:

$$\mathbf{w}_1 = (\mathbf{P} \otimes \mathbf{1}_N)^{\mathrm{T}} \mathbf{A}^{\mathrm{T}} \mathbf{r}_0 \tag{5}$$

The 1-hop compromise vector $\mathbf{r}_1$ is defined as $\mathbf{r}_1 = \mathbb{H}_a(\mathbb{T}(\mathbf{w}_1))$. In effect, the operator $\mathbb{H}_a$ compares the threshold average number of walks with the hardening level of each host. This means that a host j can be compromised only when the thresholded average number of 1-hop walk $[\mathbb{T}(\mathbf{w}_1)]_j$ is greater than its hardening level $[\mathbf{a}]_j$. Generalizing this result to h-hop, we have:

$$\mathbf{w}_{h+1} = (\mathbf{P} \otimes \mathbf{1}_N)^{\mathrm{T}} \mathbf{A}^{\mathrm{T}} \mathbf{w}_h \tag{6}$$

$$\mathbf{r}_{t+1} = \mathbb{H}_a \left(\mathbb{T} \left(\sum_{h=1}^{t+1} \mathbf{w}_h \right) \right) \tag{7}$$

The term in Eq. (7) has an equivalent expression:

$$\mathbf{r}_{t+1} = \mathbb{H}_a \left(\mathbb{T} \left(\mathbf{r}_t + (\mathbf{P} \otimes \mathbf{1}_N)^{\mathrm{T}} \mathbf{A}^{\mathrm{T}} \mathbf{r}_t \right) \right) \tag{8}$$

Thus, the matrix $\mathbf{J} = (\mathbf{P} \otimes \mathbf{1}_N)^{\mathrm{T}} \mathbf{A}^{\mathrm{T}}$ is the dominant factor for lateral movement on the host-application graph, and Eq. (8) leads to an iterative algorithm for reachability computation in host-application graphs.

Reachability of Lateral Movement on Tripartite User-Host-Application Graph

Utilizing the results from sections "Reachability of Lateral Movement on User-Host Graph" and "Reachability of Lateral Movement on Host-Application Graph", the cascading process of lateral movement on the tripartite user-host-application graph can be modeled by:

$$\mathbf{r}_{t+1} = \mathbb{H}_a\left(\mathbb{T}\left(\mathbf{r}_t + \left[\mathbf{B} + (\mathbf{P} \otimes \mathbf{1}_N)^\mathrm{T}\mathbf{A}^\mathrm{T}\mathbf{r}_t\right]\right)\right)$$

Segmentation on User-Host Graph

In this section, we investigate segmentation on user-host graphs as a countermeasure for suppressing lateral movement. Segmentation works by creating new user accounts to separate user from host in order to reduce the reachability of lateral movement, as illustrated in Fig. 2b. In principle, segmentation removes some edges from the access graph G_C and then merges these removed edges to create new user accounts. Therefore, segmentation retains the same access functionality and constrains lateral movement attacks at the price of additional user accounts. The following analysis provides a theoretical framework of different segmentation strategies.

Recall from Eq. (3) that the matrix $\mathbf{B}$ is the key factor affecting the reachability of lateral movement on G_C. Therefore, an effective edge removal approach for segmentation is reducing the spectral radius of $\mathbf{B}$, that is, $\lambda_{\max}(\mathbf{B})$ by reducing some edges from G_C. Note that, by definition, $\mathbf{B} = \mathbf{A}_C^\mathrm{T}\mathbf{A}_C$, so that $\mathbf{B}$ is a positive semidefinite (PSD) matrix and all entries of $\mathbf{B}$ are non-negative. Thus by the Perron-Frobenius theorem [23], the entries of $\mathbf{B}$'s largest eigenvector $\mathbf{u}$ (i.e., the eigenvector such that $\mathbf{Bu} = \lambda_{\max}(\mathbf{B})\mathbf{u}$) are non-negative.

Here we investigate the change in $\lambda_{\max}(\mathbf{B})$ when an edge is removed from G_C in order to define an edge score function that is associated with spectral radius reduction of $\mathbf{B}$. If an edge $(i,j) \in \mathcal{E}(\$i,j)$ is removed from G_C, then the resulting adjacency matrix of $G_C \setminus (i,j)$ is $\tilde{\mathbf{A}}_C((i,j)) = \mathbf{A}_C - \mathbf{e}_i^U\mathbf{e}_j^{N^\mathrm{T}}$. The corresponding induced adjacency matrix is:

$$\tilde{\mathbf{B}}((i,j)) = \tilde{\mathbf{A}}_C((i,j))^\mathrm{T}\tilde{\mathbf{A}}_C((i,j))$$
$$= \mathbf{B} - \mathbf{A}_C^\mathrm{T}\mathbf{e}_i^U\mathbf{e}_j^{N^\mathrm{T}} - \mathbf{e}_j^N\mathbf{e}_i^{U^\mathrm{T}}\mathbf{A}_C + \mathbf{e}_j^N\mathbf{e}_j^{N^\mathrm{T}} \tag{9}$$

By the Courant-Fischer theorem [23], we have:

$$\lambda_{\max}\left(\tilde{\mathbf{B}}\left((i,j)\right)\right) \geq \mathbf{u}^{\mathsf{T}}\tilde{\mathbf{B}}\left((i,j)\right)\mathbf{u} = \lambda_{\max}(\mathbf{B}) - 2\mathbf{u}^{\mathsf{T}}\mathbf{A}_C^{\mathsf{T}}\mathbf{e}_i^U[\mathbf{u}]_j + [\mathbf{u}]_j^2 \quad (10)$$

The relation in Eq. (10) leads to a greedy removal strategy that finds the edge $(i,j) \in \mathcal{E}$, which maximizes the edge score function $2\mathbf{u}^{\mathsf{T}}\mathbf{A}_C^{\mathsf{T}}\mathbf{e}_i^U[\mathbf{u}]_j - [\mathbf{u}]_j^2$, in order to minimize a lower bound on the spectral radius of $\mathbf{B}(i,j)$. Let $f(\mathcal{E})$ denote the function that provides a score that evaluates the effect of edge removal set $\mathcal{E}(\mathcal{R})$ on the spectral radius of $\tilde{\mathbf{B}}(\mathcal{E}_\mathcal{R})$. The function is given by:

$$f(\mathcal{E}_\mathcal{R}) = 2\sum_{(i,j)\in\mathcal{E}_\mathcal{R}} \mathbf{u}^{\mathsf{T}}\mathbf{A}_C^{\mathsf{T}}\mathbf{e}_i^U[\mathbf{u}]_j - \sum_{i\in\mathcal{V}_{\text{user}}}\sum_{j\in\mathcal{V}_{\text{host}},\,(i,j)\in\mathcal{E}_\mathcal{R}}\sum_{s\in\mathcal{V}_{\text{host}},\,(i,s)\in\mathcal{E}_\mathcal{R}}[\mathbf{u}]_j[\mathbf{u}]_s$$

$$(11)$$

$f(\mathcal{E}_\mathcal{R})$ is non-negative as it can be represented as the sum of non-negative terms. It can be readily shown that the following property holds:

Property 1 For any edge removal set $\mathcal{E}_\mathcal{R}$ with $|\mathcal{E}_\mathcal{R}| = q \geq 1$, if there exists one edge removal set $\mathcal{E}_\mathcal{R} \subset \mathcal{E}$ such that $f(\mathcal{E}_\mathcal{R}) > 0$, then there exists some constant c such that:

$$\lambda_{\max}(\mathbf{B}) - c.f(\mathcal{E}_\mathcal{R}) \geq \lambda_{\max}\left(\tilde{\mathbf{B}}\left(\mathcal{E}_\mathcal{R}\right)\right) \quad (12)$$

Based on Property 1, we can deduce that the edge score function $2\mathbf{u}^{\mathsf{T}}\mathbf{A}_C^{\mathsf{T}}\mathbf{e}_i^U[\mathbf{u}]_j - [\mathbf{u}]_j^2$ is associated with an upper bound on the spectral radius $\tilde{\mathbf{B}}\left((i,j)\right)$, $f(\mathcal{E}_\mathcal{R})$ is associated wih an upper bound on the spectral radius of $\tilde{\mathbf{B}}(\mathcal{E}_\mathcal{R})$, and maximizing $f(\mathcal{E}_\mathcal{R})$ is an effective strategy for spectral radius reduction of $\mathbf{B}$. Further, $f(\mathcal{E}_\mathcal{R})$ is a monotonic increasing set function, which means that for any two subsets $\mathcal{E}_{\mathcal{R}_1}, \mathcal{E}_{\mathcal{R}_2} \subset \mathcal{E}$ satisfying $\mathcal{E}_{\mathcal{R}_1} \subset \mathcal{E}_{\mathcal{R}_2} f(\mathcal{E}_{\mathcal{R}_2}) \geq f(\mathcal{E}_{\mathcal{R}_1})$. Also, $f(\mathcal{E}_\mathcal{R})$ is a monotone submodular set function [24]. Submodularity means that for any $\mathcal{E}_{\mathcal{R}_1} \subset \mathcal{E}_{\mathcal{R}_2} \subset \mathcal{E}$ and edge $e \in \mathcal{E}\backslash\mathcal{E}_{\mathcal{R}_2}$, the discrete derivative $\Delta f(e|\mathcal{E}_\mathcal{R}) = f(\mathcal{E}_\mathcal{R} \cup e) - f(\mathcal{E}_\mathcal{R})$ satisfies $\Delta f(e|\mathcal{E}_{\mathcal{R}_2}) \leq \Delta f(e|\mathcal{E}_{\mathcal{R}_1})$. This establishes a performance guarantee of greedy edge removal on reducing the spectral radius of $\mathbf{B}(\mathcal{E}_\mathcal{R})$.

Algorithm 1, given in Fig. 3, is a greedy segmentation algorithm that computes the edge score function $f((i,j)) = 2\mathbf{u}^{\mathsf{T}}\mathbf{A}_C^{\mathsf{T}}\mathbf{e}_i^U[\mathbf{u}]_j - [\mathbf{u}]_j^2$ for every edge $(i,j) \in \mathcal{E}$ and segments q edges of highest scores to create new user accounts.

Property 2 Let $\mathcal{E}_\mathcal{R}^{\text{opt}}$ be the optimal batch edge removal set with $\left|\mathcal{E}_\mathcal{R}^{\text{opt}}\right| = q \geq 1$ that maximizes $f(\mathcal{E}_\mathcal{R})$. Let also $\mathcal{E}_\mathcal{R}^q$ with $\left|\mathcal{E}_\mathcal{R}^q\right| = q$ be the greedy edge removal set obtained from Algorithm 1. If $f\left(\mathcal{E}_\mathcal{R}^q\right) > 0$, then there exists some constant $c' > 0$ such that:

$$f\left(\mathcal{E}_\mathcal{R}^{\text{opt}}\right) - f\left(\mathcal{E}_\mathcal{R}^q\right) \leq \left(1 - \frac{1}{q}\right)^q f\left(\mathcal{E}_\mathcal{R}^{\text{opt}}\right) \leq \frac{1}{e}f\left(\mathcal{E}_\mathcal{R}^{\text{opt}}\right) \quad (a)$$

Algorithm 1 Greedy score segmentation algorithm

Input: $\mathbf{A}_C$, number of segmented edges q
Output: modified access adjacency matrix $\mathbf{A}_C^q$
if recalculating score **then**

    Initialization: $\mathbf{A}_C^{old} = \mathbf{A}_C$. $\mathcal{E}_{old} \leftarrow \mathcal{E}$. $\mathcal{E}_{\mathcal{R}} \leftarrow \varnothing$.
    for $z = 1$ to q **do**
        1. Compute the leading eigenvector $\mathbf{u}$ of
            $\mathbf{B} = \mathbf{A}_C^{old\,T} \mathbf{A}_C^{old}$
        2. Compute score $f\left((i,j)\right) = 2\mathbf{u}^T \mathbf{A}_C^{old\,T} \mathbf{e}_i^U [\mathbf{u}]_j$
            $- [\mathbf{u}]_j^2$ for all $(i,j) \in \mathcal{E}_{old}$
        3. Remove the highest scored edge $(i^*, j^*) \in \mathcal{E}_{old}$
            from $\mathbf{A}_C^{old}$
        4. $\mathbf{A}_C^{old} = \mathbf{A}_C^{old} - \mathbf{e}_{i^*}^U \mathbf{e}_{j^*}^{N\,T}$. $\mathcal{E}_{old} \leftarrow \mathcal{E}_{old} \setminus (i^*, j^*)$.
            $\mathcal{E}_{\mathcal{R}} \leftarrow \mathcal{E}_{\mathcal{R}} \cup (i^*, j^*)$.

    else
        1. Compute the leading eigenvector $\mathbf{u}$ of $\mathbf{B} = \mathbf{A}_C^T \mathbf{A}_C$
        2. Compute score $f\left((i,j)\right) = 2\mathbf{u}^T \mathbf{A}_C^T \mathbf{e}_i^U [\mathbf{u}]_j - [\mathbf{u}]_j^2$
            for all $(i,j) \in \mathcal{E}$
        3. Remove the q edges of highest scores from $\mathbf{A}_C$
        4. Store this set of q edges in $\mathcal{E}_{\mathcal{R}}$

    5. Segment the removed edges in $\mathcal{E}_{\mathcal{R}}$ to create new users.
    A new user u has access to a set of hosts $\{s : (u, s) \in \mathcal{E}_{\mathcal{R}}\}$
    6. Obtain the modified access adjacency matrix $\mathbf{A}_C^q$

Fig. 3 Greedy score segmentation algorithm (with score recalculation)

$$\lambda_{\max}(\mathbf{B}) - f\left(\mathcal{E}_{\mathcal{R}}^{\text{opt}}\right) \leq \lambda_{\max}\left(\tilde{\mathbf{B}}\left(\mathcal{E}_{\mathcal{R}}^q\right)\right) \leq \lambda_{\max}(\mathbf{B}) - c'f\left(\mathcal{E}_{\mathcal{R}}^{\text{opt}}\right) \tag{b}$$

Property 3 Let $\tilde{\mathbf{A}}(\mathcal{E}_{\mathcal{R}})$ denote the adjacency matrix of $G_C \setminus \mathcal{E}_{\mathcal{R}}$ for some $\mathcal{E}_{\mathcal{R}} \subset \mathcal{E}$, and let $\mathbf{u}_{\mathcal{E}_{\mathcal{R}}}$ denote the largest eigenvector of $\tilde{\mathbf{B}}(\mathcal{E}_{\mathcal{R}})$. For any edge removal set $\mathcal{E}_{\mathcal{R}} \subset \mathcal{E}$, let $f_{\mathcal{E}_{\mathcal{R}}}(i, j) = 2\mathbf{u}_{\mathcal{E}_{\mathcal{R}}}^T \tilde{\mathbf{A}}(\mathcal{E}_{\mathcal{R}})^T \mathbf{e}_i^U [\mathbf{u}_{\mathcal{E}_{\mathcal{R}}}]_j - [\mathbf{u}_{\mathcal{E}_{\mathcal{R}}}]_j^2$. Let also, (i^*, j^*) be a maximizer of $f_{\mathcal{E}_{\mathcal{R}}}(i, j)$. Then $\lambda_{\max}\left(\tilde{\mathbf{B}}(\mathcal{E}_{\mathcal{R}})\right) \geq \lambda_{\max}\left(\tilde{\mathbf{B}}(\mathcal{E}_{\mathcal{R}} \cup (i^*, j^*))\right)$. Furthermore, if $f_{\mathcal{E}_{\mathcal{R}}}(i^*, j^*) > 0$, then $\lambda_{\max}\left(\tilde{\mathbf{B}}(\mathcal{E}_{\mathcal{R}})\right) > \lambda_{\max}\left(\tilde{\mathbf{B}}(\mathcal{E}_{\mathcal{R}} \cup (i^*, j^*))\right)$.

Properties 2 and 3 basically tell us that score recalculation can successively reduce the spectral radius of $\mathbf{B}$ and ensures that Algorithm 1 has a better performance guarantee on spectral radius relative to the optimal batch edge removal strategy of combinatorial computation for selecting the best q edges.

Property 4 Let $\mathbf{d}^U = \mathbf{A}_C \mathbf{1}_N$ and $\mathbf{d}^N = \mathbf{A}_C^T \mathbf{1}_U$ denote the degree vectors of users and hosts, respectively, and the terms $d_{\max}^{\text{user}}$ and $d_{\max}^{\text{host}}$ denote the maximum degree of users and hosts in G_C. If an edge (i, j) is removed from G_C, and $\tilde{\mathbf{B}}(i, j)$ is irreducible, then:

Algorithm 2 Greedy user-(host-)first segmentation algorithm

Input: $\mathbf{A}_C$, number of segmented edges q
Output: modified access adjacency matrix $\mathbf{A}_C^q$
Initialization: $\mathbf{A}_C^{old} = \mathbf{A}_C$. $\mathcal{E}_{old} \leftarrow \mathcal{E}$. $\mathcal{E}_\mathcal{R} \leftarrow \varnothing$.
for $z = 1$ to q **do**
 1. Compute user (host) degree vector $\mathbf{d}^U = \mathbf{A}_C^{old}\mathbf{1}_N$
  $(\mathbf{d}^N = \mathbf{A}_C^{old^T}\mathbf{1}_U)$
 2. Obtain $i^* = \arg\max_i [\mathbf{d}^U]_i$ and
  $j^* = \arg\max_{j:[\mathbf{A}_C^{old}]_{i^*j}>0}[\mathbf{d}^N]_h$
  $(j^* = \arg\max_j[\mathbf{d}^N]_j$ and
  $i^* = \arg\max_{i:[\mathbf{A}_C^{old}]_{ij^*}>0}[\mathbf{d}^U]_i)$
 3. Remove the edge $(i^*, j^*) \in \mathcal{E}_{old}$ from $\mathbf{A}_C^{old}$.
 4. $\mathbf{A}_C^{old} = \mathbf{A}_C^{old} - \mathbf{e}_{i^*}^U \mathbf{e}_{j^*}^{N^T}$. $\mathcal{E}_{old} \leftarrow \mathcal{E}_{old} \setminus (i^*, j^*)$.
  $\mathcal{E}_\mathcal{R} \leftarrow \mathcal{E}_\mathcal{R} \cup (i^*, j^*)$
 5. Segment the removed edges in $\mathcal{E}_\mathcal{R}$ to create new users.
 A new user u has access to a set of hosts $\{s : (u, s) \in \mathcal{E}_\mathcal{R}\}$
 6. Obtain the modified access adjacency matrix $\mathbf{A}_C^q$

Fig. 4 Greedy user-(host)-first segmentation algorithm

$$\lambda_{\max}\left(\tilde{\mathbf{B}}(i, j)\right) \leq d_{\max}^{\text{user}} \cdot d_{\max}^{\text{host}} - \max_{s \in \{1, 2, \ldots, N\}}\left[\left([\mathbf{d}^U]_i - 1\right)\mathbf{e}_j^N - \mathbf{A}_C^T\mathbf{e}_i^U\right]_s \quad \text{(c)}$$

Property 4 suggests a greedy user-(host)-first segmentation approach that segments the edge between the user with a maximum degree, while the corresponding accessible host of maximum degree in order to reduce the upper bound on spectral radius (in Property 4). Figure 4 presents a greedy user-(host)-first segmentation algorithm based on Property 4.

Hardening on Host-Application Graph

In this section, we discuss two countermeasures for constraining lateral movements on the host-application graph. Edge-hardening refers to securing access from application k to host j and in effect reducing the compromise probability $[\mathbf{P}]_{kj}$. Node-hardening refers to securing a particular host and, in effect, increasing its hardening level. Recall from Eq. (8) that the reachability of lateral movement on the host-application graph is governed by the matrix $\mathbf{J} = (\mathbf{P} \otimes \mathbf{1}_N)^T\mathbf{A}^T$. Although $\mathbf{J}$ is in general not a symmetric matrix, it is matrix of non-negative entries and, hence, by the Perron-Frobenius theorem [23], $\lambda_{\max}(\mathbf{J})$ is real and non-negative, and the entries of its largest eigenvector are non-negative.

Hardening a host j for an application k means that after hardening, the compromise probability $[\mathbf{P}]_{kj}$ is reduced to some value ϵ_{kj} $[\mathbf{P}]_{kj} > \epsilon_{kj} \geq 0$. Let $\mathcal{H}$ denote the set of hardened edges, and let $\mathbf{P}_{\mathcal{H}}$ be the compromise probability matrix after edge-hardening.   Then   we   have $\tilde{\mathbf{P}}_{\mathcal{H}} = \mathbf{P} - \sum\limits_{(k,j) \in \mathcal{H}} \left([\mathbf{P}]_{kj} - \epsilon_{kj} \right) \mathbf{e}_k^K \mathbf{e}_j^{N^{\mathrm{T}}}$.   Let

$\widetilde{\mathbf{J}}(\mathcal{H}) = \left(\tilde{\mathbf{P}}_{\mathcal{H}} \otimes \mathbf{1}_N \right)^{\mathrm{T}} \mathbf{A}^{\mathrm{T}}$, and let $\mathbf{y}$ be the largest eigenvector of $\mathbf{J}$. We can show that:

$$\lambda_{\max} \left(\widetilde{\mathbf{J}}(\mathcal{H}) \right) \geq \lambda_{\max}(\mathbf{J}) - \mathbf{y}^{\mathrm{T}} \Delta \mathbf{J}_{\mathcal{H}} \mathbf{y} \tag{13}$$

$$\Delta \mathbf{J}_{\mathcal{H}} = \left[\left(\sum_{(k,j) \in \mathcal{H}} \left([\mathbf{P}]_{kj} - \epsilon_{kj} \right) \mathbf{e}_k^K \mathbf{e}_j^{N^{\mathrm{T}}} \right) \otimes \mathbf{1}_N \right]^{\mathrm{T}} \mathbf{A}^{\mathrm{T}} \tag{14}$$

Let $\phi(\mathcal{H}) = \mathbf{y}^{\mathrm{T}} \Delta \mathbf{J}_{\mathcal{H}} \mathbf{y}$ be a score function that reflects the effect of the edge-hardening set $\mathcal{H}$ on spectral radius reduction of J. It can be easily shown that $\phi(\mathcal{H})$ is a monotonic increasing set function of $\mathcal{H}$.

Figure 5 provides a greedy edge-hardening algorithm (with score recalculation and without) that hardens the η edges of highest scores between applications and hosts, where the per-edge-hardening score is defined as $\phi((k,j)) = \mathbf{y}^{\mathrm{T}} \Delta \mathbf{J}_{(k,j)} \mathbf{y}$.

The following three properties hold on Algorithm 3.

Algorithm 3 Greedy edge hardening algorithm

Input: $\mathbf{J} = (\mathbf{P} \otimes \mathbf{1}_N)^T \mathbf{A}^T$, number of hardened edges η,
$\{\epsilon_{kj}\}_{k \in \{1,2,\ldots,K\}, j \in \{1,2,\ldots,N\}}$
Output: modified compromise probability matrix $\mathbf{P}^\eta$
if recalculating score **then**
    Initialization: $\mathbf{P}^\eta = \mathbf{P}$. $\mathbf{J}^{old} = \mathbf{J}$.
    for $z = 1$ to η **do**
        1. Compute the leading eigenvector $\mathbf{y}$ of $\mathbf{J}^{old}$
        2. Compute score $\phi((k,j)) = \mathbf{y}^T \Delta \mathbf{J}_{(k,j)}^{old} \mathbf{y}$
        3. Obtain $(k^*, j^*) = \arg\max_{k,j} \phi((k,j))$
        4. Edge hardening: $[\mathbf{P}^\eta]_{k^* j^*} = \epsilon_{k^* j^*}$
        5. $\mathbf{J}^{old} = (\mathbf{P}^\eta \otimes \mathbf{1}_N)^T \mathbf{A}^T$ (see Appendix P)
else
    Initialization: $\mathbf{P}^\eta = \mathbf{P}$
    1. Compute the leading eigenvector $\mathbf{y}$ of $\mathbf{J}$
    2. Compute score $\phi((k,j)) = \mathbf{y}^T \Delta \mathbf{J}_{(k,j)} \mathbf{y}$
    3. Find the η edges of highest scores
    4. Store this set of η edges in $\mathcal{H}$
    5. Edge hardening: $[\mathbf{P}^\eta]_{kj} = \epsilon_{kj}$ for all $(k,j) \in \mathcal{H}$

Fig. 5 Greedy edge-hardening algorithm

Algorithm 4 Greedy node hardening algorithm

Input: edge score $\phi((k,j))$, number of hardened nodes ζ, $\{\alpha_j\}_{j=1}^N$

Output: modified node hardening vector $\tilde{\mathbf{a}}$

Initialization: $\tilde{\mathbf{a}} = \mathbf{a}$

1. Compute edge hardening score $\phi((k,j))$ for all $k \in \{1, 2, \ldots, K\}$ and $j \in \{1, 2, \ldots, N\}$
2. Compute node hardening score $\rho(j) = \sum_{k=1}^K \phi((k,j))$ for all $j \in \{1, 2, \ldots, N\}$
3. Find the first ζ nodes of highest scores and store this set of ζ nodes in $\mathcal{H}^{node}$
4. Node hardening: $[\tilde{\mathbf{a}}]_j = \alpha_j$ for all $j \in \mathcal{H}^{node}$

Fig. 6 Greedy nodel hardening algorithm

Property 5 For any hardening set $\mathcal{H}$ with $|\mathcal{H}| = \eta \geq 1$, let $\mathcal{H}^\eta$ with $|\mathcal{H}^\eta| = \eta$ be the greedy hardening set obtained from Algorithm 3. Then $\mathcal{H}^\eta$ is a maximizer of $\phi(\mathcal{H})$.

Property 6 For any hardening set $\mathcal{H}$ with $|\mathcal{H}| = \eta \geq 1$, $\lambda_{\max}(\mathbf{J}) \geq \lambda_{\max}\left(\widetilde{\mathbf{J}}(\mathcal{H})\right)$. Furthermore, let $\mathcal{H}^{\text{opt}}$ with $|\mathcal{H}^{\text{opt}}| = \eta$ be the optimal hardening set that minimizes $\lambda_{\max}\left(\widetilde{\mathbf{J}}(\mathcal{H})\right)$, and let $\mathcal{H}^\eta$ with $|\mathcal{H}^\eta| = \eta$ be the hardening set that maximizes $\phi(\mathcal{H})$. If $\lambda_{\max}(\mathbf{J}) > 0$ and $\phi(\mathcal{H}^\eta) > 0$, then there exists some constant $c'' > 0$ such that $\lambda_{\max}(\mathbf{J}) - \phi(\mathcal{H}^\eta) \leq \lambda_{\max}\left(\widetilde{\mathbf{J}}(\mathcal{H}^{\text{opt}})\right) \leq \lambda_{\max}(\mathbf{J}) - c''\phi(\mathcal{H}^\eta)$.

Property 7 Let $\mathbf{y}_\mathcal{H}$ denote the largest eigenvector of $\widetilde{\mathbf{J}}(\mathcal{H})$, and let $\phi_\mathcal{H}((k,j)) = \mathbf{y}_\mathcal{H}^{\mathrm{T}} \widetilde{\mathbf{J}}(\mathcal{H} \cup (k,j))\mathbf{y}_\mathcal{H}$. For any edge-hardening set $\mathcal{H}$, let (k^*, j^*) be a maximizer of $\phi_\mathcal{H}((k,j))$. Then $\lambda_{\max}\left(\widetilde{\mathbf{J}}(\mathcal{H})\right) \geq \lambda_{\max}\left(\widetilde{\mathbf{J}}(\mathcal{H} \cup (k^*, j^*))\right)$. Furthermore, if $\lambda_{\max}\left(\widetilde{\mathbf{J}}(\mathcal{H})\right) > 0$, then $\lambda_{\max}\left(\widetilde{\mathbf{J}}(\mathcal{H})\right) > \lambda_{\max}\left(\widetilde{\mathbf{J}}(\mathcal{H} \cup (k^*, j^*))\right)$. Thus, Algorithm 3 with score recalculation can successively reduce the spectral radius of $\mathbf{J}$.

For the last algorithm on hardening, we use the edge-hardening score $\phi_\mathcal{H}((k,j))$ to define the node-hardening score $\rho(j)$ for host j, where $\rho(j) = \sum_{k=1}^K \phi((k,j))$. In effect, node-hardening on host j enhances its hardening level from $[\mathbf{a}]_j$ to a value $\alpha_j \in [[\mathbf{a}]_j, 1]$. A greedy algorithm based on the node-hardening score is given in Fig. 6.

Experimental Results

Dataset Description and Experiment Setup

To demonstrate the effectiveness of the proposed segmentation and hardening strategies against lateral movement attacks, we use the event logs and network flows collected from a large enterprise to create a tripartite user-host-application graph as in Fig. 2a for performance analysis. This graph contains 5863 users, 4474 hosts, 3 applications, 8413 user-host access records, and 6230 host-application-host network flows. All experiments assume that the defender has no knowledge of which nodes are compromised, and the defender only uses the given tripartite network configuration for segmentation and hardening.

To simulate a lateral movement attack, we randomly select five hosts (approximates 0.1% of total host number) as the initially compromised hosts and use the algorithms developed in section "Network Model and Iterative Reachability Computation of Lateral Movement" to evaluate the reachability, which is defined as the fraction of reachable hosts by propagating on the tripartite graph from the initially compromised hosts. The initial node-hardening level of each host is independently and uniformly drawn from the unit interval between 0 and 1. The compromise probability matrix $\mathbf{P}$ is a random matrix where the fraction of nonzero entries is set to be 10% and each nonzero entry is independently and uniformly drawn from the unit internal between 0 and 1. The compromise probability after hardening, ε_{kj}, is set to be 10^{-5} for all k and j. All experimental results are averaged over ten trials.

Segmentation Against Lateral Movement

Figure 7 shows the effect of different segmentation strategies proposed in section "Segmentation on User-Host Graph" on the user-host graph. In particular, Fig. 7a presents the results of reachability with respect to different segmentation strategies, while Fig. 7b shows the fraction of newly created user accounts from segmentation. The results in Fig. 7a clearly show that given the same number of segmented edges, a greedy host-first segmentation strategy is the most effective approach to constraining reachability at the cost of most additional accounts, since accesses to high-connectivity hosts (i.e., hubs) are segmented. For example, segmenting 15% of user-host accesses can reduce the reachability to nearly one-third of its initial value. Greedy segmentation with score recalculation is shown to be more effective than that without score recalculation since it is adaptive to user-host access modification during segmentation. Greedy user-first segmentation strategy is not as effective as the other strategies since segmentation does not enforce any user-host access reduction, and therefore, after segmentation, a user can still access the hosts but with different accounts.

Fig. 7 Effect of
segmentation on the user-
host access graph

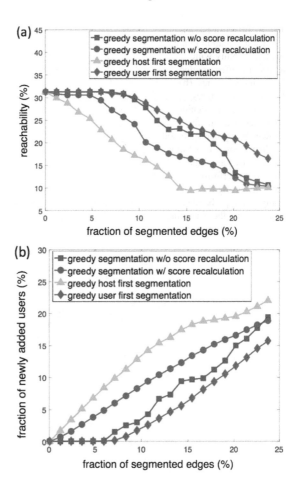

Figure 7b shows the fraction of newly created accounts with respect to different segmentation strategies. There is clearly a trade-off between network security and implementation practicality since Fig. 7b suggests that segmentation strategies with better reachability reduction capability also lead to more additional accounts. However, in practice, a user might be reluctant to use many accounts to pursue his/her daily jobs even though doing so can greatly mitigate the risk from lateral movement attacks.

Hardening Against Lateral Movement

Figure 8 shows the effect of different hardening strategies proposed in section "Hardening on Host-Application Graph" on the host-application graph. Specifically, Fig. 8a presents the results of reachability with respect to different edge-hardening

Fig. 8 Effect of hardening
on host-application graph.
(a) Reachability with
respect to different edge-
hardening strategies. (b)
Reachability with respect to
different node-hardening
strategies. The greedy
hardening approaches based
on the proposed hardening
matrix **J** (red and blue
curves) outperform
heuristics using the
compromise probability
matrix **P** and the hardening
level vector **a** (green curve)

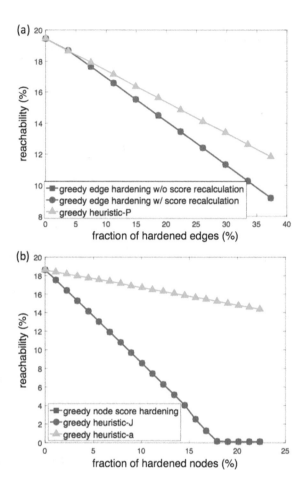

strategies, while Fig. 8b shows the results of reachability with respect to different node-hardening strategies. As shown in Fig. 8a, the proposed greedy edge-hardening strategies with and without score recalculation have similar performance in reachability reduction, and they outperform the greedy heuristic strategy that hardens edges of the highest compromise probability. This suggests that the proposed edge-hardening strategies indeed find the nontrivial edges affecting lateral movement. Figure 8b shows that the node-hardening strategies using the node score function ρ and $\rho^{\mathbf{J}}$ lead to similar performance in reachability reduction, and they outperform the greedy heuristic strategy, which harden the nodes of the lowest hardening level.

These results show that the greedy edge- and node-hardening approaches based on the proposed hardening matrix **J** outperform heuristics using the compromise probability matrix **P** and the hardening level vector **a**, which suggest that the intuition of hardening the host of lowest security level might not be the best strategy for constraining lateral movement, as it does not take into account the connectivity structure of the host-application graph.

Performance Evaluation on Actual Lateral Movement Attacks

This section demonstrates the importance of incorporating the heterogeneity of a cyber-system for enhancing the resilience to lateral movement attacks. Specifically, real lateral movement attacks taking place in an enterprise network are collected as a performance benchmark.[1] This dataset contains the communication patterns between 2010 hosts via two communication protocols, and therefore, the enterprise network can be summarized as a bipartite host-application graph. It also contains lateral movements originated from a single compromised host and in total includes 2001 propagation paths.

This benchmark dataset was collected from the network traffic of a cyber-testbed running inside an OpenStack-based cloud with nearly 2000 virtual machine instances. Starting from a known machine (host), the attack involved logging from one machine to another using SSH. Implemented by automated scripts, on each machine the attack replicated to four other machines at the beginning of every hour. This process continued for 8 hours. We collected network traffic flows from each virtual machine and combined to produce a 16 GB packet capture dataset. Each packet information was further aggregated to produce "flow"-level information, which can be interpreted as a "communication session" between two machines. As an example, when a client connects to the server, the client may send five packets and receive ten packets of data from the server. The "flow" level data will combine these 15 data packets into a single "flow" to represent one interaction between the machines. Each flow record has the following elements: IP address and port information for both source and destination devices, protocol, flow start time, duration, and message size. We infer the application by considering the protocol and destination address pair. As an example, a flow to destination port 22 over TCP implies an SSH connection. To apply our proposed method to the cyber-system against lateral movement attacks, we select the source and destination IP address and the applications to build the host-application graph.

The experiment in this section differs from the analysis in section "Experimental Results", as this dataset contains actual lateral movement traces on the host-application graph, whereas in section "Experimental Results", we have a complete user-host-application tripartite graph of an enterprise but without the actual attack traces.

We compare the performance of our proposed edge-hardening method (see Algorithm 3) to the NetMelt algorithm [1, 22], which is a well-known edge-removal method for containing information diffusion on a homogeneous graph. For the proposed edge-hardening method, the edges in the host-application bipartite graph are hardened sequentially according to the computed scores, and the initial compromise probability matrix **P** is set to be a matrix of ones. For every propagation

---

[1]Dataset available at https://sites.google.com/site/pinyuchenpage/datasets

Fig. 9 Performance evaluation on the collected dataset

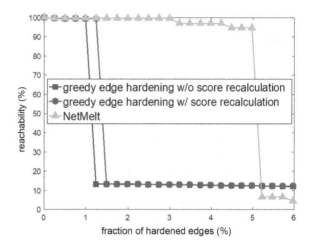

path, the lateral movement will be contained if the edge it attempts to leverage is hardened. Since NetMelt can only deal with homogeneous graphs (in this case, the host-host graph), its recommendation on hardening a host pair is equivalent to hardening K corresponding host-application edges (in this case, $K = 2$), whereas our method has better granularity for edge-hardening by considering the connectivity structure of the host-application bipartite graph. The computation complexity of NetMelt is $O(m\eta + N)$, where m is the number of edges in the host-host graph, η is the number of hardened edges, and N is the number of hosts. Since the operation of leading eigenpair computation in Algorithm 3 is similar to NetMelt, the computation complexity for Algorithm 3 without score recalculation is $O(m'\eta + N)$, where m' is the number of nonzero entries in the matrix $\mathbf{J}$. For Algorithm 3 with score recalculation, the computation complexity is $O(m'\eta^2 + N\eta)$.

Figure 9 shows the reachability of lateral movements with respect to the fraction of hardened edges. Initially, the reachability is nearly 100%, suggesting that almost every host is vulnerable to lateral movement attacks without edge-hardening. It can be observed that the proposed method (both with or without score recalculation) can restrain the reachability to roughly 10% by hardening less than 1.5% of edges, whereas NetMelt requires to harden more than 5% of edges to achieve comparable reachability, since it does not exploit the heterogeneity of the cyber-system. Consequently, the results demonstrate the utility of incorporating heterogeneity for building resilient systems.

Conclusion and Future Work

The information presented in this chapter developed a framework for joint modeling of multiple dimensions of cyber-behavior (user access control, application traffic) for enhancing cyber-enterprise-resiliency in a unified, tripartite network model. Our experiments performed on a real dataset demonstrate the value and powerful insights from this unified model with respect to analysis performed on a single dimensional dataset. Through the tripartite graph model, the dominant factors affecting lateral movement are identified, and effective algorithms are proposed to constrain the reachability with theoretical performance guarantees. We also synthesized a benchmark dataset containing traces of actual lateral movement attacks. The results showed that our proposed approach can effectively contain lateral movements by incorporating the heterogeneity of the cyber-system. Our future work includes generalization to k-partite networks to model other dimensions of behavior (e.g., authentication mechanisms and social profile of users).

References

1. H. Tong, B.A. Prakash, T. Eliassi-Rad, M. Faloutsos, C. Faloutsos, Gelling and melting, large graphs by edge manipulation, in *Proceedings of the 2012 ACM International Conference on Information and Knowledge Management*, pp. 245–254, 2012, https://doi.org/10.1145/2396761.2396795.
2. H. Goldman, R. McQuaid, J. Picciotto, Cyber-resilience for mission assurance, in *Proceedings of the 2011 IEEE International Conference on Technologies for Homeland Security (HST)*, pp. 236–241, 2011, https://doi.org/10.1109/THS.2011.6107877.
3. N. Provos, M. Friedl, P. Honeyman, Preventing privilege escalation, in *Proceedings of the 2003 USENIX Security Symposium*, 2003.
4. S. Bugiel, L. Davi, A. Dmitrienko, T. Fischer, A.-R. Sadeghi, B. Shastry, Towards taming privilege-escalation attacks on android, in Proceedings of the 2012 Network and Distributed Systems Security, p. 19, 2012.
5. P.-Y. Chen, C.-C. Lin, S.-M. Cheng, H.-C. Hsiao, C.-Y. Huang, Decapitation via digital epidemics: A bio-inspired transmissive attack. IEEE Commun. Mag. **54**, 75–81 (2016). https://doi.org/10.1109/MCOM.2016.7497770
6. L. Xing, X. Pan, R. Wang, K. Yuan, X. Wang, Upgrading your Android, elevating my malware: privilege escalation through mobile OS updating, in *Proceedings of the 2014 IEEE Symposium on Security and Privacy*, pp. 393–408, 2014, https://doi.org/10.1109/SP.2014.32
7. T. Das, R. Bhagwan, P. Naldurg, Baaz: a system for detecting access control misconfigurations, in *Proceedings of the 2010 USENIX Security Symposium*, pp. 161–176, 2010.
8. Y. Chen, S. Nyemba, B. Malin, Detecting anomalous insiders in collaborative information systems. IEEE Trans. Dependable Secure Comput. **9**, 332–344 (2012). https://doi.org/10.1109/TDSC.2012.11
9. A. Zheng, J. Dunagan, A. Kapoor, Active graph reachability reduction for network security and software engineering. IJCAI Artif. Intell. J. **22**, 1750 (2011)
10. P.-Y. Chen, S.-M. Cheng, K.-C. Chen, Optimal control of epidemic information dissemination over networks. IEEE Trans. Cybern. **44**, 2316–2328 (2014). https://doi.org/10.1109/TCYB.2014.2306781
11. S.-M. Cheng, W.C. Ao, P.-Y. Chen, K.-C. Chen, On modeling malware propagation in generalized social networks. IEEE Commun. Lett. **15**, 25–27 (2011). https://doi.org/10.1109/LCOMM.2010.01.100830
12. A. Chapman, M. Nabi-Abdolyousefi, M. Mesbahi, Controllability and observability of network-of-networks via Cartesian products. IEEE Trans. Autom. Control **59**, 2668–2679 (2014). https://doi.org/10.1109/TAC.2014.2328757
13. J. Gao, S.V. Buldyrev, S. Havlin, H.E. Stanley, Robustness of a network of networks. Phys. Rev. Lett. **107**(195701) (2011). https://doi.org/10.1103/PhysRevLett.107.195701

14. J. Ni, H. Tong, W. Fan, X. Zhang, Inside the atoms: ranking on a network of networks, in *Proceedings of the 2014 ACM SIGKDD International Conference on Knowledge Discovery and Data Mining*, pp. 1356–1365, 2014, https://doi.org/10.1145/2623330.2623643.

15. M. Halappanavar, S. Choudhury, E. Hogan, P. Hui, J. Johnson, I. Ray, L. Holder, Towards a network-of-networks framework for cybersecurity, in *Proceedings of the IEEE Intelligence and Security Informatics Conference*, pp. 106–108, 2013.

16. P. Demeester, M. Gryseels, A. Autenrieth, C. Brianza, L. Castagna, G. Signorelli, R. Clemenfe, M. Ravera, A. Jajszczyk, D. Janukowicz, K.V. Doorselaere, Y. Harada, Resilience in multilayer networks. IEEE Commun. Mag. **37**, 70–76 (1999). https://doi.org/10.1109/35.783128

17. S. Choudhury, P.-Y. Chen, L. Rodriguez, D. Curtis, P. Nordquist, I. Ray, K. Oler, Action recommendation for cyber-resilience, in *Proceedings of the 2015 Workshop on Automated Decision Making for Active Cyber-Defense*, pp. 3–8, 2015, https://doi.org/10.1145/2809826.2809837.

18. H. Chan, L. Akoglu, H. Tong, Make it or break it: manipulating robustness in large networks, in *Proceedings of the 2014 SIAM International Conference on Data Mining*, pp. 325–333, 2014, https://doi.org/10.1137/1.9781611973440.37.

19. P.-Y. Chen, A.O. Hero, Assessing and safeguarding network resilience to nodal attacks. IEEE Commun. Mag. **52**, 138–143 (2014). https://doi.org/10.1109/MCOM.2014.6957154

20. L.A. Adamic, C. Faloutsos, T.J. Iwashyna, B.A. Prakash, H. Tong, Fractional immunization in networks, in *Proceedings of the Siam International Conference on Data Mining*, pp. 659–667, 2013, https://doi.org/10.1137/1.9781611972832.73.

21. P. Hu, W.C. Lau, How to leak a 100-million-node social graph in just one week – a reflection on OAuth and API design in online. Soc. Networks (2014)

22. L.T. Le, T. Eliassi-Rad, H. Tong, MET: A fast algorithm for minimizing propagation in large graphs with small Eigen-gaps, in *Proceedings of the 2015 SIAM International Conference on Data Mining*, pp. 694–702, 2015, https://doi.org/10.1137/1.9781611974010.78.

23. R.A. Horn, C.R. Johnson, *Matrix Analysis* (Cambridge University Press, New York, NY, 1990)

24. S. Fujishige, *Submodular Functions and Optimization: Annals of Discrete Math* (North Holland, Amsterdam, 1990)

Part III
Proactive Defense Mechanism Design

Moving Target, Deception, and Other Adaptive Defenses

Benjamin Blakely, William Horsthemke, Alec Poczatec, Lovie Nowak, and Nathaniel Evans

Abstract Moving target defenses raise the cost of an attack to make it more difficult or infeasible. Strategies to do so include implementing diversity, movement, and obfuscation at the platform, network, runtime environment, application, or data layer. Doing so, however, often requires an investment in software, hardware, procedure, or overhead (such as training) and can also increase the complexity of infrastructures being defended. In industrial control system contexts, this complexity and its impact upon performance and reliability might present obstacles to implement such defensive technologies. As the scope of possible domains for introducing moving target defense concepts is now well-defined and the considerations are largely enumerated, consideration must additionally be given to systems that can dynamically select optimal strategies in response to attacks. In this chapter, we will survey the foundations, principles, and domains of moving target defense, consider specific implementation examples, and evaluate the considerations for implementing deceptive and responsive strategies in industrial control systems applications.

The work presented in this paper was partially supported by the US Department of Energy, Office of Science under DOE contract number DE AC02-06CH11357. The submitted manuscript has been created by UChicago Argonne, LLC, operator of Argonne National Laboratory. Argonne, a DOE Office of Science laboratory, is operated under Contract No. DE-AC02-06CH11357. The US Government retains for itself, and others acting on its behalf, a paid-up nonexclusive, irrevocable worldwide license in said article to reproduce, prepare derivative works, distribute copies to the public, and perform publicly and display publicly, by or on behalf of the government.

B. Blakely · W. Horsthemke · A. Poczatec · L. Nowak · N. Evans (✉)
Strategic Security Sciences Division, Argonne National Laboratory, Lemont, IL, USA
e-mail: bblakely@anl.gov; horst@anl.gov; poczatec@anl.gov; lnowak@anl.gov; nevans@anl.gov

© Springer Nature Switzerland AG 2019

C. Rieger et al. (eds.), *Industrial Control Systems Security and Resiliency*, Advances in Information Security 75, https://doi.org/10.1007/978-3-030-18214-4_6

Introduction

It is widely acknowledged that cyberspace is an asymmetric domain. Defenders must constantly expend resources to evaluate threats and risks, deploy mitigating strategies, and watch for the inevitable attempt at a compromise. Being wrong once, missing even a single patch, or failing to close one critical port is the digital equivalent of leaving the safe door open and unattended. As documented in the media, there are no shortage of examples to make this point, and to attempt to enumerate them would be both reductive and redundant. Attackers have many advantages. Primary among these are a virtually unbounded time horizon to conduct reconnaissance and carry out an attack and the fact that little more than a keen intellect and an Internet connection might be needed to cause significant damage. While the imbalance in terms of timing is as old as conflict itself, this ability to cause great damage with few resources is largely unique.

Nowhere is this more evident than in industrial control system (ICS) applications. The repercussions for a cyberattack in many sectors can be very expensive and harmful to the livelihoods and happiness of individuals on a scale of a few to millions. However, it is particular to the ICS sector that the consequences for these attacks can threaten safety, and even lives, on a similar scale. ICS are used in many applications, from process control and factory automation to controlling regional water distribution systems and international electrical grids. The asymmetry of cybersecurity is perhaps nowhere more on display than here.

Moving target, deception, and other adaptive defenses have garnered increasing attention in the literature over the past 15 years. Technologies under these headings attempt to put a thumb on the other end of the cybersecurity scale and, if not prevent, at least raise the cost of a cyberattack. If a cyber-system can throw virtual sand in the eyes of its attacker—morphing and modulating proactively or in response to an ongoing attack—it may buy time for detection and mitigation. Even knowing that a system is employing such tactics might be enough to deter an opportunistic cybercriminal.

However, in an ICS context, the constraints upon performance and reliability are paramount. These defenses all introduce complexity and may impact the ability of a system to complete its function within required parameters or to maintain compatibility with other components. ICSs require high levels of assurance, and thus any new technologies introduced must be thoroughly understood, tested, and verified in a broad range of environments. Initial work on moving target defenses has shown success for ICS, but work remains to be done.

The remainder of this chapter is laid out as follows: Section "Foundations of Moving Target Defense" provides additional background on the concepts of moving target defenses, including their historical antecedents and ideal properties. Section "Types of Moving Target Defense" presents a categorical classification of the layers at which moving target defenses may be implemented. Section "Examples of Moving Target Defense" gives additional detail for this classification, giving specific examples from the literature of defenses fitting each category.

Section "Industrial Control System Applications of Moving Target Defense" places this work within the frame of ICS considerations and shows examples of where moving target defenses have been developed for that context. Section "Strategy Selection" considers the problem of responsiveness; that is, how can we choose defenses a priori or enable systems to make real-time decisions to maximize effectiveness against likely threats. Section "Conclusions" presents conclusions.

Foundations of Moving Target Defense

Moving target defenses build upon principles developed for altering deterrence calculus in a missile defense context. Specifically with systems and software engineering, many of the concepts overlap those with an objective of fault tolerance. A cyberattack can be considered as an adversarial action or as a fault. From the purposes of resilience, security can be impacted by non-intentional actions to a similar degree as by a human attacker. Thus, the general problem may be framed as how to make a system maximally resilient to threats—both foreseen and unforeseen. This requires deterring threats from coming to fruition (in the case of an adversary), minimizing the chance of success and scope of damage, and maximizing the chances of detection.

Missile defense Much of the history of moving target defenses can be traced to missile defense during the Cold War. Both sides attempted to maximize their survivability and thus the credibility of the deterrent threat. Through the nuclear triad (i.e., intercontinental ballistic missiles [ICBMs], submarine-launched ballistic missiles [SLBMs], and strategic bombers), the goal was to ensure that the enemy could not entirely disable the retaliatory capability in one fell swoop. Making it as difficult as possible to locate missiles naturally led to making the launch platforms themselves mobile.

This is not to be confused with research involved with moving target countermeasures, which dealt with the challenge of identifying, tracking, and targeting a high-speed moving target (i.e., aircraft or missile). This concept is considered in Boyell [1]. Specifically addressed is the kinetic relationship among three entities: (1) a missile, representative of the attacking party; (2) a ship, representing the moving target; and (3) an antimissile, which represents the counter-weapon that the moving target uses against the attacker. This relationship is utilized in the calculation of detection range and reaction time, thereby determining the requirements for the successful defense of a ship by launching antimissiles.

The former concept, and the one more relevant to the discussions of this paper, is discussed in Sagan [2]. As a method of decreasing ICBM vulnerability, the Soviet Union constructed ICBM systems on its railways and roadways to make them moving targets and thus less susceptible to an attack. Similarly proposed, but never approved, were two moving target designs in the United States (US): (1) a rail-garrison missile-experimental (MX) plan, which would have placed 50 MX missiles onto railway cars and moved them from air force bases to commercial railroads during incidents, and (2) a single-warhead small ICBM (SICBM) system, which would have scattered SICBMs by making them ground-mobile as well. Perhaps due to the overlap in expertise and personnel, these concepts have been extended by the US Department of Defense (DoD), in particular to defend cyber-infrastructures.

Byzantine fault tolerance In many ways, moving target defenses are an expression of fault tolerance concepts. A system designed to be impervious to naturally occurring faults must be able to detect and correct anomalous operations. Likewise, a system designed for resilience to cyberattackers must exhibit predictable behavior in the face of an attack, either self-correcting or allowing ample time for a human operator to intercede.

Lamport et al. [3] address fault tolerance as the Byzantine Generals Problem. Byzantine army generals (analogous to computer processors) are imagined to camp on opposite sides, outside of an enemy city, and wish to decide upon a plan of action. However, they can communicate only by messenger. A number of the generals might be traitors and thus attempt to confuse the loyal generals. Just as in many cyber-conflicts, there is not only a lack of information but potentially erroneous information. The loyal generals must have an algorithm that satisfies two conditions: (1) all loyal generals agree upon a single plan of action (or input value), and (2) traitors cannot provoke loyal generals to elect an unsuccessful plan. The algorithm put forth in this paper is shown to be able to satisfy such conditions.

In Castro and Liskov [4], a state machine replication algorithm for Byzantine fault tolerance (BFT) is presented. It defines BFS, the first Byzantine-fault-tolerant file system. With symmetric cryptography for message authentication, BFS guarantees that even during a denial-of-service (DoS) attack, the service never returns a bad reply. Instead, it delays replies until a valid one can be made. Additionally, the underlying BFT library guarantees the integrity of invariants and access controls in the presence of faulty clients. A replicated system can tolerate faults over its lifetime, as long as less than one-third of the replicas are compromised within a vulnerability period. Here, we see the application of fault tolerance concepts directly to a security challenge. While the system is static, the existence of replicas is similar to a number of proposed moving target defenses.

The concept of moving targets is applied to BFT and crash fault tolerant (CFT) hybrid protocols in Duan et al. [5]. The approach switches between the two protocols based on a given system's consensus-protocol performance, as well as vulnerabilities within the network, which are assessed by Intrusion Detection System (IDS) signals. As a cost-conscious approach, it accounts for damage cost (protocol vulnerability) and operational cost (cost of running the protocol) in order to predict which replicas may be faulty before running them.

N-version system design The concept of diverse redundancy is well-established in system engineering literature. N-version fault tolerance is introduced in Avizienis [6]. The paper presents several issues regarding its implementation, including initial specification, independence of design efforts, N-version execution support, protection of the support environment, architectural support, recovery of failed versions, modification of N-version software, assessment of effectiveness, and cost. Three components of multiple computation as it applies to the N-version approach are discussed as well—time, space, and information.

Laprie et al. [7] extend these ideas by classifying various approaches to software and hardware fault tolerance, including N-version programming (NVP), recovery block (RB) programming, and N self-checking programming (NSCP). All of these approaches create an array of computational units (like the Byzantine Generals above) that are as independent as possible and thus unlikely to fall victim to the same attacks or exhibit the same vulnerabilities. These are again static approaches but valuable concepts that will be seen through moving target defense research.

Voting, an essential aspect of N-version programming, is addressed in Parhami [8]. Presented voting algorithms are comprised of four parts: input data, output data, input vote, and output vote. These four parts are used in a further classification scheme: (1) voting-type exact, inexact, or approval, (2) consensus vs. compromise, (3) output-selection rule threshold or plurality, (4) preset vs. adaptive, and (5) input object space properties, size, and structure. Notably, as a precursor to formal definitions for evaluating moving target defenses, this paper also offers theorem proofs for designers of fault-tolerant systems.

A firewall design is presented in Liu and Gouda [9], inspired by the concept of design diversity in N-version programming. The approach divides into three phases: (1) design, in which many versions of the firewall policy are created; (2) comparison, in which the versions are compared in order to find incompatibilities between them; and (3) resolution, in which the inconsistencies are resolved and a firewall is created upon which all creators agree. In order to discover the incompatibilities, the paper offers construction, shaping, and comparison algorithms. This practical implication looks very much like some of the network-based moving target defenses developed since.

Moving Target Defense Principles

Moving target defenses, in a cyber-context, incorporate many principles of the previously discussed concepts. Like missile defense, a cyber-defender is in a constant feedback loop of deterrence calculus and must signal to would-be attackers that it is not in their best interest to come after this target or, if they do, to ensure that they must contend with additional ambiguity and unknowns. The Byzantine Generals scenario is analogous to many types of cyberattacks whereby not only does the defender not know whether an attack is imminent or ongoing, but information needed to make such determinations may also not be reliable. As in N-version system design, redundancy of functionality with diversity as a core component can both increase the effort required to understand a system and compartmentalize the damage that can be done by a single attack.

Moving target defenses thus seek to capture such properties as components of system design. In Okhravi et al. [10] (after a definition by the US Networking and Information Technology Research and Development Program [NITRD]), the primary objectives of these defenses are defined as making the system less homogeneous, less static, and less deterministic. Zhuang et al. [11] express a similar idea in the concepts of configuration space, diversification, and randomization.

Diversity Making the system less homogeneous, that is, more diverse (or a broader configuration space), means intentionally introducing an array of divergent applications and systems. For example, a simple website may traditionally be served by a single system or an array of systems of identical configuration. Applying diversity to this environment might entail rotating incoming requests among systems with different operating systems, web server daemons, application frameworks, or programming languages. Attackers are highly dependent upon system vulnerabilities to find their way into a target.

While it has been repeatedly proven that no system is immune from such defects, the more dissimilar two systems are, the less likely they are to share the same defect. Accordingly, an attack vector may allow for compromise of a single system but require further work to achieve lateral movement within the infrastructure. Carter et al. [12], discussed further in Responsive Defenses, consider the relative contributions of diversity and randomness, finding that diversity is a critical component of a moving target defense strategy.

In reality, it can be difficult to achieve true diversity as many systems share common libraries. CVE-2014-0160 (Heartbleed) was perhaps the first widely publicized example of this [13]. The library underpinning encryption routines for a vast number of systems of many different configurations and manufacturers contained a serious defect that allowed an attacker to remotely extract confidential information from other users of a website. The extent to which such a fundamental library was distributed in the cyber-ecosystem was a key lesson learned from this event (and unfortunately has been relearned several times since). Foreseeing the extent to which two systems of apparently different types overlap in "vulnerability space" can be very challenging. It is therefore important, in introducing diversity, to focus on the most fundamental layers (even the hardware itself) or to use a mixed approach combining multiple layers.

Movement Even a system with many different and divergent components can still be enumerated by a determined attacker if the configuration is static. If the system profile (or attack surface, the system properties relevant to enabling an attack) does not change, there is nothing to stop an attack from identifying a weak point in the system and exploiting it. Movement, whether proactive (i.e., a normal period of rotation or reconfiguration) or reactive (i.e., system modifications in response to an ongoing attack attempt), requires the attacker to constantly shift tactics.

In the context of a typical attack sequence, this means that the steps of reconnaissance and enumeration must be repeated continuously every time the configuration changes. Typically, these steps are completed once (or perhaps periodically but infrequently in a long-term campaign), and the attacker then focuses on using this information to build and execute an attack strategy. Of course, information gathering never truly stops, and an attacker exploiting a target may continue to gather new information that refines the understanding of the target. But if the system is constantly in movement, this is not a matter of refinement but complete refactoring of the attack strategy.

Randomization At the most advanced level, an attacker might be able to learn enough about a system to determine the existent system permutations and movement schedule. Thus, it is important that moving target defenses incorporate randomness in their strategies. An attacker should not be able to make reliable predictions about a system state at the time of the attack.

This can be a very difficult feature to incorporate fully. The feasibility of presenting a fully randomized configuration upon each service request is low. More tractable is a set of diversified configurations prepared in advance and presented to attackers on a random schedule. However, features such as Address Space Layout Randomization (ASR/ASLR) [14], incorporate randomness in a more fundamental way. Cryptographers know well the difficulty of obtaining truly random input and its impact upon security [15]. Given enough resources, an attacker may be able to reduce an insufficiently random rotation schedule to become essentially nonrandom.

Types of Moving Target Defense

Moving target defenses can be implemented in a number of ways, at different levels of a system's architecture. In general, the further "down the stack" (i.e., at a more fundamental level) these defenses can be applied, the greater impact they will have upon an adversary. However, this is not true in all cases, and layering multiple moving target defenses might afford additional benefits. Regardless, for an analysis of examples of existing moving target defenses, we will use the four-layer model from Okhravi et al. [10] and consider defenses at the layers of network, platform, runtime environment, application, and data.

Moving target methods help mitigate persistent threats and confuse the methods used to collect information about the attack surface. Moving target methods range from migrating applications between different types of systems with diverse operating systems, libraries, and architectures to rotating between different variants of an application, such as a web service, that uses different frameworks, layouts, and support systems. Other moving target methods assume that attacks will compromise their system or application and continually refresh their system with a clean, trusted version.

Network-based Methods for dynamically changing networks may include changing addresses, changing topology, or changing the protocols used for communication. Dynamic network strategies range from constantly changing the addresses and ports of the essential systems, obscuring information about the host and protocols by hiding or randomizing the packet headers, creating overlay networks to increase the resiliency of traffic flows, and using voting systems to determine whether to trust the integrity of the traffic. While these may be implemented without awareness by the applications themselves, an intermediate layer, whether at the operating system or network level, must be employed to track the changes to ensure uninterrupted connectivity.

Platform-based The category of dynamic platforms includes methods that change various properties of the platform, such as the central processing unit (CPU), operating system, architecture, or instances. The dynamic platforms category contains a variety of strategies, including simultaneously employing multiple diverse variants of the application, moving between different types of platforms, and enhancing the isolation or protection of the application.

Diversity strategies may deploy dynamic runtime environment methods, such as system call mapping, unpredictable stack directions, and instruction set tagging. These methods make it much more difficult for an attacker to exploit flaws in applications running on the platform that would rely upon information regarding the state of the operating system.

Runtime environment-based In characterizing dynamic runtime environments, Okhravi further described methods that randomize either the address space (the layout of memory) or the instruction set (interfaces used by applications to interact with the process or the input/output devices) [10]. This method might also be described as modifying the system software. Changing the runtime environment is an attempt to prevent attackers from exploiting knowledge they have gained or assumed about the runtime environment. These methods attempt to mitigate injection attacks, which attempt to execute malicious code or change the control of program execution.

ASLR decreases the likelihood that the attacker can predict the addresses used for code control and execution. Instruction set randomization (ISR) decreases the likelihood that the system will execute code injected by an attacker. The methods work by altering how the system interprets the code used for the instructions. The alterations range from tagging to encrypting the code. Because the system understands only the altered code, the system will not recognize and/or execute code injected by an attacker. These techniques can be deployed without any awareness by the applications themselves and thus are often the least complicated to implement.

Application-based Methods for dynamically changing the application software consist of modifying instructions, including the order, grouping, and format. Defensive strategies for deploying dynamic software tend to employ multiple versions of the application operating in diverse ways. One method aims to improve overall application resiliency by deploying multiple diverse versions to increase the likelihood that at least one version will survive a system-wide attack.

Another approach enhances this resilient-availability method by enhancing the overall integrity though the use of application voting. Each application uses a different version of a dynamic runtime environment, which increases the likelihood that at least one will withstand each individual attack. When an application request arrives, each application receives the request and submits a response (a vote). If a majority of the responses agree, the integrity of the response is confirmed and the response is delivered.

A third method focuses on detecting attacks, creating two versions of an application. One version grows the program execution upward in memory space and one downward. Because attacks, such as buffer overflows, tend to exploit position on the stack, attacks would typically not succeed on both types of stack management methods. If an application monitor detects differences in behavior, this indicates that an attack is occurring.

Data-based Methods for dynamically changing data focus on techniques that alter how the application expects to handle data. These techniques include changing the format, syntax, encoding, or representation of the application data or randomization techniques. The latter method randomizes all code and data using random keys at runtime. This prevents attackers from injecting code or controls. It simultaneously maintains data confidentiality by randomizing all data read by an attacker.

Various dynamic data methods employ the use of multiple versions of the application to detect and reject malicious input. Monitors analyze data as it flows through multiple versions of the same application to detect variations that indicate malicious code. The implementation of multiple versions include modifying the application to transform the data differently as well as running the application in different address spaces or varying the instruction set used in data transformation.

Voting methods are also employed for dynamically defending against malicious input. Multiple independent versions of the application receive the data and output their results to a voting method that determines whether the output is acceptable. These methods sometimes approximate the data, but if exact data is used, the voting method will select the majority output.

Examples of Moving Target Defense

Platform-Based

Roeder and Schneider [16] describe a method of proactive obfuscation to create a diverse set of servers to reduce the incidence of shared vulnerabilities. To increase server diversity, they modify the executable codes to create different versions of the daemon. The obfuscation method preserves the semantics of the program but creates diverse executables through variations in compilation, loading, or runtime execution. This diversity aims to increase the work required by adversaries.

Carvalho et al. [17] described a moving target defense command and control framework (MTC2) to enhance human decision-making by incorporating intelligent agents and moving target defenses. Both humans and intelligent agents can decide how and when to employ the moving target defense. To implement the moving target defense, they employ random redirection of traffic to a diverse set of applications. Carvalho et al. [18] describe Mission-aware Infrastructure for Resilient Agents (MIRA), a cyber-command and control infrastructure that integrates human decision-makers with cyber-agents and moving target defenses that can continually change and adapt to ever-changing adversaries. Carvalho et al. [19] describe a semiautomated approach for wrapping dynamic and moving target cyber-defenses (SAWD) for cyber-command and control operations. The wrapper aims to create a package that upon execution will install/uninstall, configure, and control a sensor, actuator, or defender. The package control interfaces allow it to be managed by command and control infrastructures.

Thompson et al. [20] describe the Multiple Operating System Rotational Environment MTD (MORE MTD), which rotates the operating system that serves the application service, typically a web-based application. The application service is given a static Internet Protocol (IP) address, and the rotation method randomly reassigns this IP address to different application operating system configurations. This strategy attempts to minimize the time an intruder can attack the system, as well as confuse attempts to harvest information about the configuration of the system.

Ahmed and Bhargava [21] describe Mayflies, a technique that continuously changes the configurations of the virtual machines that form their distributed infrastructure. By creating diverse, randomized variants of virtual machine-based applications that operate for only short-time intervals then vanish, their moving target defense stops in-progress attacks and limits the spread of attacks. In addition to removing applications after they finish tasks at specific time intervals, they also terminate applications if they detect integrity violations. They proactively monitor system state by comparing runtime memory with a separate trusted reference memory. If this virtual introspection detects an anomaly, the application is terminated and a new variant replaces it. This may be of particular interest to software engineering teams, which operate in dynamically scaling or continuous integration infrastructures, as some of these benefits may be realized unintentionally.

Network-Based

In a method that looks much like the Byzantine fault tolerance methods above, Ho et al. [22] present Nysiad, a method that tolerates not only crashes but also uncertain, inconsistent process behaviors. To handle these potential faults, Nysiad assigns several guard hosts to each process. The guards replicate and verify the messages handled by the processes. This method increases fault tolerance at the cost of substantial replication and overhead.

In Jafarian et al. [23], IP addresses are randomized to deter, deceive, and detect attackers. Their method assigns ephemeral IP addresses to each host and constrains connectivity between these ephemeral IP addresses by space and time. Jafarian et al. [24] also propose using proxy honeypots to constantly and randomly mutate addresses and fingerprints of hosts to try to anonymize their identity and thwart adversarial reconnaissance.

Macfarland and Shue [25] present an example of a technique, the SDN Shuffle, using software-defined networking (SDN), which is a technology that virtualizes large parts of the network infrastructure and allows for a high degree of intelligence over decisions regarding network routing. This may therefore enable a large degree of network-level motion and diversity with little integration needed with the systems and applications running on top. SDN Shuffle periodically replaces actual IP and media access control (MAC) addresses with synthetic addresses to thwart reconnaissance. Attackers must then expend extra effort tracking which systems have which addresses, cast a broader net (which will be much more noisy), or repeat steps of their attack whenever the environment shifts.

Skowyra et al. [26] propose a randomization method to hide network identifiers and encrypt the data payloads. Their method uses OpenFlow-based SDN to replace IP addresses with temporary nonces. By obfuscating the packet headers from the MAC layer and above, they aim to reduce the accuracy of adversarial attempts to identify and characterize the systems.

Runtime Environment-Based

Kc et al. [27] developed an ISR method to create process-specific methods for executing potentially malicious software. Their method requires a key to execute the ISR algorithm. Because an attacker is unlikely to know the key, the code they attempt to inject will not run and will instead cause a runtime exception. Their method considers both compiled executables and scripting and interpreted languages, such as web-based SQL injections.

Shacham et al. [14] explore the effectiveness of address space randomization and conclude that randomization is ineffective on 32-bit machines. Compile-time randomization works better than runtime methods because it creates finer-grained addresses, but compile-time methods are more vulnerable to information leakage through examining their binary codes. ASLR is a feature present in nearly all modern operating systems and, while it has been shown to be less than perfect [28], drastically raises the difficulty of exploiting what are otherwise the most popular types of software vulnerabilities. The impact to applications is little to none, and applications may not even be aware of its presence. For these reasons, it is the single most successful embodiment of moving target defense principles to date.

Application-Based

Huang and Ghosh [29] rotate a set of diversified web servers to decrease the predictability of the attack surface. The diversification method affects several layers of the software stack: (1) application layer, (2) web server layer, (3) operating system layer, and (4) virtualization layer. The rotation can occur either periodically or if the system detects anomalies or fails integrity checks. Similarly, Thompson et al. [30] dynamically redirect web-based traffic to different types of web-based applications. By redirecting traffic at random, this method increases the cost of adversarial reconnaissance and decreases the likelihood that an attacker, if successful in finger-printing a web server, can exploit any vulnerability.

In contrast to introducing diversity to the web daemon, Vikram et al. [31] devised a method called NOMAD, which uses randomization of HTML elements to limit or prevent the web bots from identifying the HTML elements used by humans when interacting with the website. This limits or prevents the web bot from automatically behaving like legitimate users.

Obfuscating the application code is the aim of two other works. Jangda et al. [32] describe an application diversification method for Java-based applications. Their method frequently regenerates the code dynamically using a Java bytecode Just-In-Time (JIT) compiler. By changing the code frequently, their method reduces the time available to attackers to discover and exploit address-based memory exploits. Mahmood and Shila [33] consider the computational limitations of Internet of Things (IoT) devices by providing trusted application code only when the device needs to operate. When necessary, the IoT device downloads the application code and operates. After use, this code is erased. Before downloading another application code, the IoT device operates using a minimal set of trusted code. To increase the resilience of the application code, the method employs code transformations or address randomization techniques.

Industrial Control System Applications of Moving Target Defense

Consideration of the use of moving target defenses in an ICS context requires an understanding of what makes those environments different. While the examples above may be applicable in certain ICS contexts, perhaps with modification, they may also create unforeseen problems given additional requirements and restrictions. In the sections that follow, we enumerate a number of these constraints and give examples from the literature of existing ICS-focused moving target defenses.

Considerations

ICS infrastructures are an aggregation of systems that use digital methods to monitor and control tangible assets. These components, or cyber-physical systems (CPS), connect embedded computing technologies to the physical world and are found in a

Table 1 Domains of cyber-physical system usage [34]

Domain	Function
Smart Manufacturing	Optimizing productivity in the manufacture of goods or delivery of services
Emergency Response	Handling threats against public safety and protecting nature and valuable infrastructures
Air Transportation	Operation and traffic management of aircraft systems
Critical Infrastructure	Distribution of daily life supplies such as water, electricity, gas, and oil
Health Care and Medicine	Monitoring health conditions of patients and taking necessary actions
Intelligent Transportation	Improving safety, coordination, and services in traffic management with real-time info sharing
Robotic for Service	Performing services for the welfare of humans

number of domains, as shown in Table 1. As their functions typically require precise control in terms of timing or signaling, they are highly sensitive to any factor that changes their performance characteristics. They may also be a part of an interconnected system of components, where failure to meet operational requirements could have cascading impacts to the entire infrastructure.

Special consideration must be given to the particular challenges in operating such systems. While similar to any information technology context, these factors are of particular importance in ICS infrastructures as the consequences for not meeting them can be significantly worse. Moving target defenses that are to be used in ICS deployments must be rigorously tested to ensure they will not upset the delicate nature of the systems they are defending. Gunes et al. [34] enumerate a number of these challenges:

Dependability: perform required functionalities during its operation without significant degradation in its performance and outcome.
Sustainability: capable of enduring without compromising requirements of the system while renewing the systems resources and using them efficiently.
Security: control access to system resources and protect sensitive information from unauthorized disclosures.
Reliability: degree of correctness, which a system provides to perform its function.
Interoperability: ability of the systems/components to work together, exchange information, and use this information to provide specified services.
Predictability: degree of foreseeing of a system's state/behavior/functionality, either qualitatively or quantitatively.

Yan et al. [35] present a similar list of factors for smart grids, an area of particular research focus within the ICS domain. The difficulty of balancing supply and demand, routing power where it needs to be, and incorporating distributed energy resources into electrical grids has required increasing amounts of automation and information sharing between grid components. Of particular note, it also specifically lists complexity as a challenge. Electrical grids are highly dynamic, and the many interoperating components make predictability difficult. Moving target defenses are likely to contribute greatly to this challenge rather than help.

For example, moving target defenses may significantly decrease predictability by design. While this makes the job of an attacker more difficult, it also increases the difficulty of guaranteeing the system will perform as required. Similarly, interoperability may be impacted if components that depend upon the defended system are not capable of dealing with the ambiguous or changing interface. At a system level, guarantees of correctness and sustainability may be predicated upon a static and well-defined system configuration.

Introducing dynamic factors into that environment may render performance guarantees null. This can be of particular concern in environments that typically rely heavily upon manufacturer support and manufacturers who have strict limitations around changes that can be made before support is no longer provided. Davidson and Andel [36] specifically analyzed a number of system-level and network-level moving target defenses in an ICS context (SCADA)[1] and found that several were likely to interfere with requirements for determinism. Others were less impactful but required further analysis. The only defense endorsed was address space randomization as described in Runtime Environment-Based. Accordingly, the best strategy may be to work with manufacturers to build moving target defenses into their products.

ICS Examples

Groat et al. [37] propose using the IPv6 protocol to rotate the current address space of hosted systems as a moving target defense against attackers. IPv6 is a natural fit for smart grid applications due to its expansive address space, giving plenty of room for the large number of devices typical to these infrastructures. Thus, rotating the address of devices in IPv6 spaces makes them extremely difficult to find, more so than in IPv4 where the address space is much smaller and scanning large segments is feasible even for attackers with few resources. Additionally, the dynamic nature of the network makes it difficult for attackers to compile information on resource usage, which may be sensitive or proprietary. An implementation of this idea, MTD6, was completed in Dunlop et al. [38] and validated the concept.

Pappa et al. [39] also make use of dynamic addressing. In this scheme, the external IP address of the gateway router (i.e., the device connecting the local network segment to the broader corporate network or Internet) is rotated while keeping the internal address the same. This has the advantage of not requiring any configuration of or awareness by internal systems. This has been tested within the PowerCyber test bed at Iowa State University with varying IP-hopping intervals to

---

[1]Supervisory Control and Data Acquisition, the network protocol used to control and monitor many ICS infrastructures.

determine the effects on throughput and delay. The delay increase was measured at 2.23 ms, and an efficient interval time was identified as 10 s to minimize the loss of throughput. The identified attacks against SCADA systems in this paper stated mitigations ranging from intervals of less than 6.5 s against address range exhaustion attacks and variable intervals to combat traffic analysis attacks.

SCADA systems can be defended using symmetric packet scheduling and an IP-hopping algorithm to manipulate both source and destination addresses within the packet as it is sent between gateway routers, as in Ulrich et al. [40] Since, like above, this strategy is implemented at the gateway routers of a given SCADA system, the implementation remains transparent to end systems. It prevents targeted attacks as both the sender and receiver gateways generate traffic that looks like each gateway has a very large number of hosts dwelling behind it. This implementation was tested with four stateless Network Address Translation (NAT) rules in effect, which allow for the packets to be properly transported with the address obfuscation intact. Testing showed initial success, though impacts on throughput are directly related to increases in rotation frequency. Additional work is needed to more fully understand the performance implications and the methods of seeding the randomization methods.

Clark et al. [41] attempt to strengthen the communication channel between operator and device. It uses a cryptographically strong and randomized key for each message input by the control system operator on a Process Control Network (PCN). The introduction of this cryptographic key for each message creates a significant obstacle for attackers and would require much more knowledge to exploit than previously identified attack methods. The required network configuration for implementation is outlined, and a case study is conducted with the proposed defense mechanism modeled by a linear quadratic regulator. The case study found that the impact of an attack can be mitigated significantly by randomizing between a small set of keys.

Strategy Selection

Given the number of methods by which moving target defense principles can be applied in a given context, a method is needed to determine which is most appropriate. This can be done in a static and a priori manner or dynamically in response to malicious activity. The former relies on typical cost-benefit analyses on the defender's part. The latter relies on automated methods to select appropriate behavior. In either case, it is important to understand the challenges to implementing moving target defenses and the methods by which their impact can be quantified.

Cost-Benefit Analysis

Existing work to formally define impacts to attacks and defenses can be found in
Zhuang et al. [42]. An interaction-based model is used to relate the target, attacker,
attack, and exploration space (i.e., the scope an attacker must enumerate and
evaluate). Each of these terms is formally defined. Mathematical models such as
these are useful for research purposes, allowing for comparison in a quantitative
manner. However, they may be difficult to implement in practice due to the requirements for input that may not be readily available (or at all) to the defender. For
example, to evaluate a defense against a particular attack in the model above, the
defender must know a fair amount about the way in which the attack will be
conducted and have a comprehensive understanding of the system being defended.

The ever-changing nature of attack techniques, tactics, and procedures, as well as
changes to the defending infrastructure, means that a static analysis may quickly
become outdated. One way to deal with the uncertainties inherent in quantifying
these parameters is to take a probabilistic approach, as in Crouse et al. [43]. Still, for
a given, well-defined attack, a prototypical environment can be used to allow for
direct comparison in a manner that might be otherwise impossible.

Hobson et al. [44] formally define three primary challenges to implementing
moving target defenses and five other considerations that may be a factor. Together,
these define the scope of both the theoretical and practical factors influencing the
development and implementation of moving target defenses in production environments. A researcher conducting analysis of moving target defenses, or working to
build a new one, is likely to be primarily concerned with what a defense must
achieve to be effective. These include:

1. *Coverage*: Exploitable elements of the surface are subject to movement, and no
   information can be learned from static components.
2. *Unpredictability*: Current and future movements cannot be predicted by the
   attacker.
3. *Timeliness*: Movement is applied between attacker observation and subsequent
   attack action.

A defender (i.e., system architect, administrator, etc.) will have additional considerations. These put the moving target defense into the greater context of the
organization. In this way, it is likely to be evaluated in the same manner as any other
component. Considerations might include:

4. *Resource overhead*: The additional requirement for processor time, memory
   space, network bandwidth, etc. must be minimized.
5. *Direct cost*: The monetary costs (e.g., development, deployment, maintenance) to
   implement the defense must be justifiable given the defensive benefits.
6. *Impact to mission*: The defended system must be able to complete its primary
   function unimpeded.
7. *Usability*: The defense must be within the capabilities of a typical defender to
   implement, monitor, and maintain.

Table 2 MTD effectiveness metrics [46]

Metric	Mission definition	Attack definition
Productivity	The rate at which mission tasks are completed	How quickly an attacker can perform and complete adversarial tasks
Success	The number of attempted mission tasks that are successfully completed	How successful an attacker may be while attempting to attack a network
Confidentiality	How much mission information is exposed to eavesdroppers, whether it can be intercepted, etc.	How much attacker activity may be visible to detection mechanisms
Integrity	How much mission information is transmitted without modification or corruption	The accuracy of information viewed by an attacker

8. *Compatibility and interoperability*: The defense must fit into the existing security and service delivery strategy in a complementary way.

A similar quantitative evaluation framework was developed in Zaffarano et al. [45] and validated in Taylor et al. [46]. Mathematical definitions are given for each of the elements in Table 2, with consideration for both defender and attacker perspectives. The model presented is conceptually straightforward but again depends on sufficient capability to measure the required inputs. Given baseline assumptions, as above, this may still allow for quantitative decision-making by those seeking to implement moving target defenses if such information is made available by the developer.

Another approach for comparing moving target defenses, using state machine models, is given in Xu et al. [47]. The contexts of evaluation are split into three tiers: (1) program, (2) system, and (3) evaluation (user interface). A program state machine can be used to evaluate individual processes on a system, with consideration for damage that may be caused by state transitions. A system state machine can consider the propagation of damage between programs (i.e., process interaction). An evaluation state machine is used to view the program and system state machines specifically with consideration for potential damage and represents a user interface to the model. Tools such as this may be used in contexts where moving target defenses, and the systems they are defending, can be decomposed and subjected to formal/static analysis.

Other Practical Considerations

A number of moving target defenses rely upon a diversity of applications, operating systems, or devices. While this inherently makes the cost of an attack higher, it may also create a burden for the defender. System administrators might be required to become familiar with a greater number of system configurations and baselines, potentially incurring additional costs for training. System development life cycles, and tooling to support them, must be adjusted or augmented to test all permutations of the system that might occur in production.

Defenders considering implementing a moving target defense must also consider how it will impact their existing processes related to compliance. Many organizations must obtain formal certification of systems prior to operation (or periodically) or in general are required to routinely satisfy the requirements of internal or external audits. A moving target defense may be more complicated to assess in a formal way, and determining how this dynamic nature maps to various standards and statutes is an area in need of further exploration.

Incident response might be more challenging in an environment implementing moving target defenses. Reconstructing the activities of an attacker, or the manner in which a system failed (whether due to malicious intent or not), requires an understanding of the system state at the time of failure. In the case of moving target defenses, it might not be possible to know that state precisely or could require additional effort to correlate log entries to make the determination. It could also increase the cost of an investigation simply through a multiplication of the number of systems or applications that need to be investigated.

Responsive Defenses

Given the variety of moving target defenses that have been developed, and the many contexts in which they might be deployed, it might be desirable to allow a system to implement the most effective defense in a given situation. This might remove the burden from a human defender of having to conduct a priori analysis and bet on a single solution and might allow the system to adapt to changing threats. On the other hand, it might further increase complexity or cost. Regardless, the matter of how a system can choose an optimal defensive strategy has been pursued from several angles.

The interplay between attackers and defenders, and how to model this relationship in a formal way, is a natural fit for game theorists. Prakash and Wellman [48] conducted a simulation of 72 game instances based on various objectives, costs, and capabilities. Through this analysis, a number of potential strategies were identified, and the importance of detection as a deterrent was highlighted in particular. When a defender cannot reliably identify attacks, moving target defenses becomes an important mitigation strategy. This mirrors the intuitive sense that a defender who is operating with incomplete information must, in essence, play the odds by making any attack as difficult as possible.

A system must be able to evaluate strategic considerations, including aspects of the threats it might encounter. Carter et al. [12] use a game theoretic approach to consider static and adaptive adversaries and develop strategies that minimize the probability of a compromise. Of note is the insight that diversity and randomization are orthogonal (and not analogous) goals. Maximizing diversity requires a fixed schedule that will minimize temporary adjacency of similar systems. Maximizing randomization inherently cannot be constrained in this way. Diversity is shown to be far superior to randomization in minimizing the probability of compromise. However, this is with a caveat that the results may be tightly coupled to the threat model used in this simulation. As with manual analysis and strategy selection, the need to consider a wide range of attacks, and the continuous evolution of them, is a critical limitation.

Rass et al. [49] use advanced persistent threats (APTs) as an example of attackers who are especially difficult to detect. APTs are generally considered to be attackers who are well-resourced and highly skilled. Such threats are difficult to detect and counter due to the diversity, novelty, and stealth of their strategies. As discussed throughout this chapter, moving target defenses are highly sensitive to changes in the attacks they were designed to counter. Rass et al. [49] address this by loosening the constraints on game objectives to be any outcome that can be ordered (i.e., ordinal) instead of specific scenarios. It accounts for the asymmetric nature of the conflict whereby a defender may know little or nothing about the attacker while the attacker has a largely unrestricted capability to probe and analyze the target to develop situational awareness. Lastly, the evaluation focuses on empirical data (versus contrived scenarios) and considers its place within an organizational risk management framework. The focus on probabilities and generalization might allow for an approach that is broadly applicable across environments, defensive strategies, and attack types.

Creating an intuitive connection between cyber-defense and game theory is a challenge for those not familiar with both topics. Lei et al. [50] provide a mapping to make this clear. They use the state-based nature of cyber-conflicts as the basis for leveraging a Markov game to drive decision-making about "hopping strategy" or how to sequence rotations. This might be a more accurate description of moving target defenses and their interaction with attackers than a matrix approach. Of particular importance is that it considers the impact to service quality from hopping behaviors in addition to the introduction of diversity. An algorithm for selecting an optimal strategy is developed and evaluated, showing that decision-making of this kind can be effectively performed by the system being defended in a responsive manner. As work of this type evolves, we expect to see products incorporating or layering multiple moving target defenses. The ability to detect an attack, characterize it, and dynamically implement an appropriate or optimal moving target defense (while not impacting service delivery) requires the coordination of multiple components and introduces complexity and risk. This is especially the case in an industrial control systems context where an inappropriate change to the infrastructure could cause cascading or catastrophic consequences. Lastly, responsive defenses depend greatly on the ability to recognize and analyze attacks automatically with a high level of confidence. This is a long-running challenge in cybersecurity research. Still, such techniques open the possibility for adaptations at the speed of attack (or greater).

Conclusion

From this overview, it is evident that research into moving target, deception, and adaptive defenses has come a long way in the past 15 years. These defenses are grounded in well-established disciplines, such as military strategy and fault tolerance. Many examples of how these principles can be incorporated in multiple domains are now available not only in literature but for production deployment. This field of interest has grown to incorporate not just software developers and system administrators but researchers who are formally defining and quantifying the properties and performance of candidate defenses. These quantitative approaches, while still difficult to be used by a defender looking to choose a strategy for implementation, have laid the foundations for objective comparisons of various techniques to determine those best-suited in specific scenarios. As new defenses are developed, it will be critical to keep practical considerations for deployment in mind: (1) cost, (2) time, (3) knowledge, and (4) complexity. By coupling quantitative methods for development and evaluation with an eye toward obstacles to operational use, researchers can maximize not only the performance but also adoption of these techniques.

Initial work in applying these techniques to ICS infrastructures has shown success. However, the many considerations of a tightly controlled and high-consequence network of cyber-physical systems must be carefully considered and will likely continue to require purpose-built moving target defenses for this domain. Layering moving target defenses on top of these systems is likely to be fraught with challenges, and thus working with manufacturers to introduce these technologies as far upstream in the development life cycle as possible is recommended.

Capabilities are now being developed to move beyond selecting appropriate defenses based on a priori cost-benefit analyses to automatically counter ongoing attacks. This research, drawing from the mature discipline of game theory, promises acceleration of defenses to the speed of attack. However, it will be critically important to ensure that such dynamic behavior is well-bounded to assure the system's ability to complete its function is impacted minimally or at least in a predictable way. At the very least, these techniques might buy time for human defenders to detect and respond before an attack is complete. All of this has certain potential to raise costs for attackers and tip the scales, if only slightly, back in favor of the defender.

References

1. R.L. Boyell, Defending a moving target against missile or torpedo attack. *IEEE Trans. Aerosp. Electron. Syst.* **AES-12**(4), 522–526 (1976). https://doi.org/10.1109/TAES.1976.308338, https://ieeexplore.ieee.org/document/4101686/
2. S.D. Sagan, Enhanced survivability and stability, in *Council on Foreign Relations*, (Princeton University Press, Princeton, 1990). https://books.google.com/books?id=d8mkO4oKWNkC

3. L. Lamport, R. Shostak, M. Pease, The Byzantine generals problem. *ACM Trans. Program. Lang. Syst.* **4**(3), 382–401 (1982). https://doi.org/10.1145/357172.357176, http://portal.acm.org/citation.cfm?doid=357172.357176, arXiv:1011.1669v3

4. M. Castro, B. Liskov, Practical byzantine fault tolerance and proactive recovery. *ACM Trans. Comput. Syst.* **20**(4), 398–461 (2002). https://doi.org/10.1145/571637.571640, http://portal.acm.org/citation.cfm?doid=571637.571640, arXiv:1203.6049v1

5. S. Duan, Y. Li, K. Levitt, Cost sensitive moving target consensus, in *Proceedings of the 15th IEEE International Symposium on Network Computing and Applications*, vol. 2, (IEEE, Cambridge, MA, 2016), pp. 272–281. https://doi.org/10.1109/NCA.2016.7778630, https://ieeexplore.ieee.org/document/7778630/

6. A. Avizienis, The N-version approach to fault-tolerant software. IEEE Trans. Softw. Eng. **SE-11**(12), 1491–1501 (1985). https://doi.org/10.1109/TSE.1985.231893, https://ieeexplore.ieee.org/document/1701972/

7. J.C. Laprie, J. Arlat, C. Beounes, K. Kanoun, Definition and analysis of hardware- and software-fault-tolerant architectures. Computer **23**(7), 39–51 (1990). https://doi.org/10.1109/2.56851, https://ieeexplore.ieee.org/document/56851/

8. B. Parhami, Voting algorithms. IEEE Trans. Reliab. **43**(4), 617–629 (1994). https://doi.org/10.1109/24.370218, https://ieeexplore.ieee.org/document/370218/

9. A.X. Liu, M.G. Gouda, Diverse firewall design. IEEE Trans. Parallel Distrib. Syst. **19**(9), 1237–1251 (2008). https://doi.org/10.1109/TPDS.2007.70802, https://ieeexplore.ieee.org/document/4384478/

10. H. Okhravi, M.A. Rabe, T.J. Mayberry, W.G. Leonard, T.R. Hobson, D. Bigelow, W.W. Streilein, *Survey of Cyber-Moving Targets* (MIT Lincoln Laboratory Technical Report, Lexington, 2013). https://www.ll.mit.edu/mission/cybersec/publications/publication-files/full-papers/2013-09-23-OkhraviH-TR-FP.pdf

11. R. Zhuang, S.A. DeLoach, X. Ou, Towards a theory of moving target defense, in *Proceedings of the First ACM Workshop on Moving Target Defense*, (ACM, Scottsdale, 2014), pp. 31–40. https://doi.org/10.1145/2663474.2663479, http://dl.acm.org/citation.cfm?doid=2663474.2663479

12. K.M. Carter, J.F. Riordan, H. Okhravi, A game theoretic approach to strategy determination for dynamic platform defenses, in *Proceedings of the First ACM Workshop on Moving Target Defense*, (ACM, New York, 2014), pp. 21–30. https://doi.org/10.1145/2663474.2663478, http://dl.acm.org/citation.cfm?doid=2663474.2663478

13. MITRE CVE-2014-0160, Heartbleed bug (2014), https://cve.mitre.org/cgi-bin/cvename.cgi?name=cve-2014-0160

14. H. Shacham, M. Page, B. Pfaff, E.J. Goh, N. Madadugo, D. Boneh, On the effectiveness of address-space randomization, in *Proceedings of the 11th ACM Conference on Computer and Communications Security*, (ACM, New York, 2004), pp. 298–307. https://doi.org/10.1145/1030083.1030124, https://dl.acm.org/citation.cfm?id=1030124

15. MITRE CWE-338, Use of Cryptographically Weak Pseudo-Random Number Generator (PRNG) (2018), https://cwe.mitre.org/data/definitions/338.html

16. T. Roeder, F.B. Schneider, Proactive obfuscation. ACM Trans. Comput. Syst. **28**(2), 1–54 (2010). https://doi.org/10.1145/1813654.1813655, http://portal.acm.org/citation.cfm?doid=1813654.1813655

17. M. Carvalho, T.C. Eskridge, L. Bunch, A. Dalton, R. Hoffman, J.M. Bradshaw, P.J. Feltovich, D. Kidwell, T. Shanklin, MTC2: A command and control framework for moving target defense and cyber-resilience, in *Proceedings of the 6th International Symposium on Resilient Control Systems*, (IEEE, San Francisco, 2013), pp. 175–180. https://doi.org/10.1109/ISRCS.2013.6623772, https://ieeexplore.ieee.org/document/6623772/

18. M. Carvalho, T.C. Eskridge, K. Ferguson-Walter, N. Paltzer, MIRA: A support infrastructure for cyber command and control operations, in *Proceedings of the 2015 Resilience Week*, (IEEE, Philadelphia, 2015), pp. 102–107. https://doi.org/10.1109/RWEEK.2015.7287426, https://ieeexplore.ieee.org/document/7287426/

19. M. Carvalho, T.C. Eskridge, M. Atighetchi, C.N. Paltzer, Semi-automated wrapping of defenses (SAWD) for cyber command and control, in *Proceedings of the IEEE Military Communications Conference 2016*, (IEEE, Baltimore, 2016), pp. 19–24. https://doi.org/10.1109/MILCOM.2016.7795295, https://ieeexplore.ieee.org/document/7795295/

20. M. Thompson, V. Kisekka, N. Evans, Multiple OS rotational environment: An implemented moving target defense, in *7th International Symposium on Resilient Control Systems*, (IEEE, Denver, 2014). https://doi.org/10.1109/ISRCS.2014.6900086, https://ieeexplore.ieee.org/document/6900086/

21. N. Ahmed, B. Bhargava, Mayflies: A moving target defense framework for distributed systems, in *Proceedings of the 2016 ACM Workshop on Moving Target Defense*, (ACM, New York, 2016), pp. 59–64, https://doi.org/10.1145/2995272.2995283, https://dl.acm.org/citation.cfm?id=2995283, arXiv:1602.05561v1

22. C. Ho, R. van Renesse, M. Bickford, D. Dolev, Nysiad: Practical protocol transformation to tolerate Byzantine failures, in *Proceedings of the 5th {USENIX} Symposium on Networked Systems Design and Implementation*, (The USENIX Association, San Francisco, 2008), pp. 175–188, https://www.usenix.org/event/nsdi08/tech/full papers/ho/ho.pdf

23. J.H.H. Jafarian, E. Al-Shaer, Q. Duan, Spatio-temporal address mutation for proactive cyber-agility against sophisticated attackers, in *Proceedings of the First ACM Workshop on Moving Target Defense*, (ACM, New York, 2014), pp. 69–78. https://doi.org/10.1145/2663474.2663483, http://dl.acm.org/citation.cfm?doid=2663474.2663483

24. J.H.H. Jafarian, A. Niakanlahiji, E. Al-Shaer, Q. Duan, Multi-dimensional host identity anonymization for defeating skilled attackers, in *Proceedings of the 2016 ACM Workshop on Moving Target Defense*, (ACM, New York, 2016), pp. 47–58. https://doi.org/10.1145/2995272.2995278, http://dl.acm.org/citation.cfm?doid=2995272.2995278

25. D.C. Macfarland, C.A. Shue, The SDN shuffle: Creating a moving-target defense using host-based software-defined networking, in *Proceedings of the Second ACM Workshop on Moving Target Defense*, (ACM, New York, 2015), pp. 37–41. https://doi.org/10.1145/2808475.2808485, https://dl.acm.org/citation.cfm?id=2808485

26. R. Skowyra, K. Bauer, V. Dedhia, H. Okhravi, Have no PHEAR: Networks without identifiers, in *Proceedings of the 2016 ACM Workshop on Moving Target Defense*, (ACM, New York, 2016), pp. 3–14. https://doi.org/10.1145/2995272.2995276, https://dl.acm.org/citation.cfm?id=2995276

27. G.S. Kc, A.D. Keromytis, V. Prevelakis, Countering code-injection attacks with instruction-set randomization, in *Proceedings of the 10th ACM conference on Computer and Communications Security*, (ACM, New York, 2003), pp. 272–280. https://doi.org/10.1145/948143.948146, http://portal.acm.org/citation.cfm?doid=948109.948146

28. H. Marco-Gisbert, I. Ripoll-Ripoll, Exploiting Linux and PaX ASLR's weaknesses on 32- and 64-bit systems, in *Black Hat Asia 2016*, (Singapore/Malaysia, 2016). https://www.blackhat.com/docs/asia-16/materials/asia-16-Marco-Gisbert-Exploiting-Linux-And-PaX-ASLRS-Weaknesses-On-32-And-64-Bit-Systems-wp.pdf

29. Y. Huang, A.K. Ghosh, Introducing diversity and uncertainty to create moving attack surfaces for web services, Chap. 8, in *Moving Target Defense*, ed. by S. Jajodia, A. K. Ghosh, V. Swarup, C. Wang, X. S. Wang, (Springer-Verlag, New York, 2011), pp. 131–159. https://doi.org/10.1007/978-1-4614-0977-9, https://www.springer.com/us/book/9781461409762

30. M. Thompson, M. Mendolla, M. Muggler, M. Ike, Dynamic application rotation environment for moving target defense, in *Proceedings of the 2016 Resilience Week*, (IEEE, Chicago, 2016), pp. 17–26. https://doi.org/10.1109/RWEEK.2016.7573301, https://ieeexplore.ieee.org/document/7573301/

31. S. Vikram, C. Yang, G. Gu, NOMAD: Towards non-intrusive moving-target defense against web-bots, in *2013 IEEE Conference on Communications and Network Security*, (IEEE, Washington, DC, 2013), pp. 55–63. https://doi.org/10.1109/CNS.2013.6682692, https://ieeexplore.ieee.org/document/6682692/

32. A. Jangda, M. Mishra, B.D. Sutter, Adaptive just-in-time code diversification, in *Proceedings of the Second ACM Workshop on Moving Target Defense*, (ACM, New York, 2015), pp. 49–53. https://doi.org/10.1145/2808475.2808487, https://dl.acm.org/citation.cfm?id=2808487

33. K. Mahmood, D.M. Shila, Moving target defense for Internet of Things using context aware code partitioning and code diversification, in *IEEE 3rd World Forum on Internet of Things*, (IEEE, Reston, 2016), pp. 329–330. https://doi.org/10.1109/WF-IoT.2016.7845457, https://ieeexplore.ieee.org/document/7845457/

34. V. Gunes, S. Peter, T. Givargis, F. Vahid, A survey on concepts, applications, and challenges in cyber-physical systems. KSII Trans. Internet Inf. Syst. **8**(12), 4242–4268 (2014). https://doi.org/10.3837/tiis.2014.12.001, http://www.itiis.org/digital-library/manuscript/894

35. Y. Yan, Y. Qian, H. Sharif, D. Tipper, A survey on smart grid communication infrastructures: Motivations, requirements, and challenges. IEEE Commun. Surv. Tutorials **15**(1), 5–20 (2013). https://doi.org/10.1109/SURV.2012.021312.00034, https://ieeexplore.ieee.org/document/6157575/

36. C. Davidson, T. Andel, Feasibility of applying moving target defensive techniques in a SCADA system, in *11th International Conference on Cyber Warfare and Security*, (Academic Conferences and Publishing International Limited, Reading, 2016). https://doi.org/10.13140/RG.2.1.5189.5441

37. S. Groat, M. Dunlop, W. Urbanksi, R. Marchany, J. Tront, Using an IPv6 moving target defense to protect the Smart Grid, in *2012 IEEE PES Innovative Smart Grid Technologies*, (IEEE, Washington, DC, 2012), pp. 1–7. https://doi.org/10.1109/ISGT.2012.6175633. https://ieeexplore.ieee.org/document/6175633/

38. M. Dunlop, S. Groat, W. Urbanski, R. Marchany, J. Tront, MT6D: A moving target IPv6 defense, in *Proceedings of the IEEE Military Communications Conference*, (IEEE, Piscataway, 2011), pp. 1321–1326. https://doi.org/10.1109/MILCOM.2011.6127486

39. A. Pappa, A. Ashok, M. Govindarasu, Moving target defense for security Smart Grid communications: Architecture, implementation, and evaluation, in *Power & Energy Society Innovative Smart Grid Technologies Conference*, (IEEE, Piscataway, 2017), pp. 3–7. https://doi.org/10.1109/ISGT.2017.8085954, https://ieeexplore.ieee.org/document/8085954/

40. J. Ulrich, J. Drahos, M. Govindarasu, A symmetric address translation approach for a network layer moving target defense to secure power grid networks, in *Proceedings of the 2017 Resilience Week*, (IEEE, Piscataway, 2017). https://doi.org/10.1109/RWEEK.2017.8088667, https://ieeexplore.ieee.org/document/8088667/

41. A. Clark, R. Poovendran, T. Basar, An impact-aware defense against Stuxnet, in *2013 American Control Conference*, (ASME, New York, 2013), pp. 4140–4147. https://doi.org/10.1109/ACC.2013.6580475, http://ieeexplore.ieee.org/lpdocs/epic03/wrapper.htm?arnumber=6580475

42. R. Zhuang, A.G. Bardas, S.A. DeLoach, X. Ou, A theory of cyber attacks a step towards analyzing MTD systems, in *Proceedings of the Second ACM Workshop on Moving Target Defense*, (ACM, New York, 2015), pp. 11–20. https://doi.org/10.1145/2808475.2808478, https://dl.acm.org/citation.cfm?id=2808478

43. M. Crouse, B. Prosser, E.W. Fulp, Probabilistic performance analysis of moving target and deception reconnaissance defenses, in *Proceedings of the Second ACM Workshop on Moving Target Defense*, (ACM, New York, 2015), pp. 21–29. https://doi.org/10.1145/2808475.2808480, http://dl.acm.org/citation.cfm?doid=2808475.2808480

44. T. Hobson, H. Okhravi, D. Bigelow, R. Rudd, W. Streilein, On the challenges of effective movement, in *Proceedings of the First ACM Workshop on Moving Target Defense*, (ACM, New York, 2014), pp. 41–50. https://doi.org/10.1145/2663474.2663480, http://dl.acm.org/citation.cfm?doid=2663474.2663480

45. K. Zaffarano, J. Taylor, S. Hamilton, A quantitative framework for moving target defense effectiveness evaluation, in *Proceedings of the Second ACM Workshop on Moving Target Defense*, (ACM, New York, 2015), pp. 3–10. https://doi.org/10.1145/2808475.2808476, http://dl.acm.org/citation.cfm?doid=2808475.2808476

46. J. Taylor, K. Zaffarano, B. Koller, C. Bancroft, J. Syversen, Automated effectiveness evaluation of moving target defenses, in *Proceedings of the 2016 ACM Workshop on Moving Target Defense*, (ACM, New York, 2016), pp. 129–134. https://doi.org/10.1145/2995272.2995282, http://dl.acm.org/citation.cfm?doid=2995272.2995282

47. J. Xu, P. Guo, M. Zhao, R.F. Erbacher, M. Zhu, P. Liu, Comparing different moving target defense techniques, in *Proceedings of the First ACM Workshop on Moving Target Defense*, (ACM, New York, 2014), pp. 97–107. https://doi.org/10.1145/2663474.2663486, http://dl.acm.org/citation.cfm?doid=2663474.2663486

48. A. Prakash, M.P. Wellman, Empirical game-theoretic analysis for moving target defense, in *Proceedings of the Second ACM Workshop on Moving Target Defense*, (ACM, New York, 2015), pp. 57–65. https://doi.org/10.1145/2808475.2808483, http://dl.acm.org/citation.cfm?doid=2808475.2808483

49. S. Rass, S. König, S. Schauer, Defending against advanced persistent threats using game-theory. PLoS One **12**(1), 1–43 (2017). https://doi.org/10.1371/journal.pone.0168675, http://journals.plos.org/plosone/article?id=10.1371/journal.pone.0168675

50. C. Lei, D.H. Ma, H.Q. Zhang, Optimal strategy selection for moving target defense based on Markov game. IEEE Access **5**, 156–169 (2017). https://doi.org/10.1109/ACCESS.2016.2633983, https://ieeexplore.ieee.org/document/7805250/

Beyond Mirages: Deception in ICS—Lessons Learned from Traditional Networks

Nate Soule and Partha Pal

Abstract Deception has been used with notable successes and failures, both offensively and defensively, in military, civilian, and personal operations for millennia. Deception for defensive use in the cyber-domain has, however, seen limited use beyond honeypots and mirages, particularly in the context of industrial control systems (ICS). In this chapter, we explore the application of deception to the defense of networked computer systems and apply recent learnings from deception in traditional systems and networks to those employed for industrial control.

Introduction

Deception has been used with notable successes [1] and failures [2], both offensively and defensively, in military, civilian, and personal operations for millennia. Deception for defensive use in the cyber-domain has, however, seen limited use beyond honeypots and mirages, particularly in the context of industrial control systems (ICS). In this chapter, we explore the application of deception to the defense of networked computer systems and apply recent learnings from deception in traditional systems and networks to those employed for industrial control.

Deception Background

The common dictionary definition of deception refers to the act of misleading by a false appearance or statement. Early work [3–5] in formulating deception as applied to the cyber-domain defines deception as the intentional distortion of one's perceived reality and situates the term in the context of a perception taxonomy—a simplified version of which is depicted in Fig. 1. While definitions abound, for the purposes of

N. Soule (✉) · P. Pal
Raytheon BBN Technologies, Cambridge, MA, USA
e-mail: nathaniel.soule@raytheon.com; partha.pal@raytheon.com

© Springer Nature Switzerland AG 2019
C. Rieger et al. (eds.), *Industrial Control Systems Security and Resiliency*, Advances in Information Security 75, https://doi.org/10.1007/978-3-030-18214-4_7

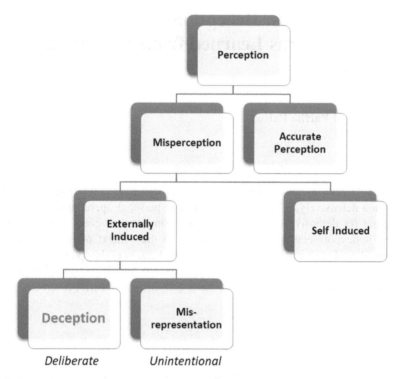

Fig. 1 Deception in the overall context of perception

this chapter, we adopt a definition that highlights a number of essential features of defensive cyber-deception:

1. It has to manipulate the adversary's perception (ideally also manipulating their decision processes).
2. It is undertaken by the defense.
3. It is done deliberately and not as an accident.

In 1991, AT&T researchers developed what may have been the first real honeypot and detailed the experience leading a particular cracker through a series of actions to attempt to learn about him, his capabilities, and his targets [6]. Concepts like honeypots, honeynets, and honey-∗ (e.g., files, ports, web pages) emerged rapidly after that and throughout the late 1990s and early 2000s. In the late 2000s, focus shifted away from pure deception to techniques that fall under the same umbrella but approach defense from a somewhat different perspective than the honey-∗ techniques, such as the moving target defense (MTD). MTDs relaxed the assumption of consistency over time and instead favored change within a system at a rate high enough such that the adversary could not keep up. Some MTDs truly embody (our definition of) deception—they alter adversarial perceptions by, for example, obfuscating the true operating system or network topology of a system/network. In these

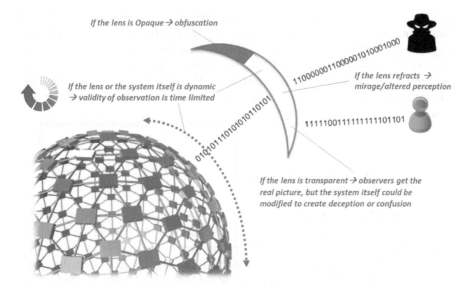

If the lens is Opaque → obfuscation

If the lens or the system itself is dynamic → validity of observation is time limited

1100000011000001010001000

If the lens refracts → mirage/altered perception

111110011111111111101101

If the lens is transparent → observers get the real picture, but the system itself could be modified to create deception or confusion

010101110101010110101

Fig. 2 Various ways to manipulate the perception of reality

cases, a static ground truth system exists if one could lift the veil, but the appearance of this system from the outside changes dynamically with time or in reaction to events. Other MTDs, in fact, do not alter perception but actually change the underlying system, again based on time- or event-based triggers. With new techniques came new descriptive terms, such as confusion and obfuscation, whose relationship with the general notion of deception can appear to be confusing at times. However, as seen in Fig. 2, these concepts can be disambiguated by decomposing the world into three entities: (1) the system itself as it exists without any deception technologies, (2) a lens through which observers perceive the system, and (3) the observers themselves. Each observer will have access to a set of observation channels based on their capabilities and access. Each of these observation channels will emit information about the underlying system. To introduce deception, one can thus change the system itself (by adding, subtracting, or modifying elements in a way that would not be done or required otherwise) or modify the data provided by the observation channels (by introducing a "lens" on the observation channels that modifies either the bits flowing through or the properties of the flows).

In all cases, the manipulation of the system or the lens can be done proactively (e.g., at start-up, based on time, randomly) or reactively (e.g., based on defender-perceived events in the system, such as detection of an adversary or of their impacts on resources).

Until recently, deception in the cyber-domain has largely been used for detection and discovery of the Tactics, Techniques, and Procedures (TTPs) and goals of an adversary. A honeypot can, for example, alert the defender to the presence of an unwanted actor (as no legitimate clients should interact with it), and monitors and probes within the honeypot can indicate how that actor is moving within the system and the methods they are using to achieve their goals. Recent work, however, has begun to expand this set of deception goals to, for example, include (1) raising the adversary's work and decision-making load and complexity, (2) prolonging the engagement to keep the adversary distracted, (3) attribution of an attacker, and (4) injection of false information to achieve collateral effects outside of the defender's network.

State of the Art in Deception in ICS

As depicted in Fig. 3, cyber-deception can be viewed as being comprised of multiple attributes. The green area is an example representation of the parts of the deception space that have seen the heaviest focus with respect to exploration and evaluation. There are certainly numerous exceptions to this focus area, but as a whole, defensive cyber-deception is still maturing. It is hard to measure and quantify its benefits, and there is a lot to explore and develop.

Among the various dimensions of deception, target environment and fidelity are of particular importance in ICS. To date, the concepts of mirages and honeypots have been the most prominently investigated/employed manifestation of deception in ICS. A honeypot is a system or pseudo-system that while it may appear to, does not provide any business function. It exists, rather, as a node in the network that no legitimate client has need to interact with, and thus all its interactions, if any, stem from malicious or misconfigured elements—making them useful for intrusion

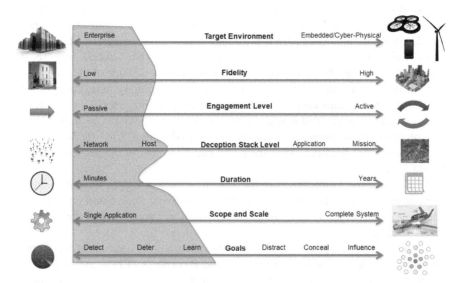

Fig. 3 Illustrative representation of the deception space

detection and adversary analysis purposes. Honeypots can range in terms of their interaction level and fidelity from systems that only present a deceptive façade and have no real depth (and thus are cheap to deploy and safer to maintain as they are not full operating nodes) to systems that truly run the operating and business services one might except from a node in the enclave, minus any sensitive data or access. The additional realism present in these "higher-fidelity" honeypots makes them potentially more effective, with higher probability of longer-duration interactions, but at the same time more difficult to build and deploy, and potentially riskier as they are in theory more likely to be compromised and used as a launching point for attacks.

As a high signal-to-noise ratio detector, honeypots in ICS have received some attention [7–10]. Deploying honeypots in industrial control settings introduces additional challenges above and beyond those present in traditional networks. Deception in the cyber-physical world involves signals/data that must obey known physical laws and act within the parameters of advertised ranges for specific hardware. While honeypots are often passive, only emitting a signal if first prompted with a request, the returned response must obey expected properties. Furthermore, the systems that deceptive ICS signals purport to emanate from are often extremely complex, employ less frequently known (compared to the traditional IT networks) and sometimes proprietary protocols, and include a wide array of devices, protocols, and formats for different vendors, ICS sectors, country of origin and usage, etc. Thus, the above-referenced work, among others, have explored mechanisms for simulating [11] or emulating real systems or automatically deriving honeypots [9] from observed traffic. Commercial deception solutions have also begun to deploy ICS-focused honeypots [12, 13] that mimic SCADA systems and act as early warning trip wires.

Mirage theory, described in Rrushi and Kang [14], takes the idea of using honeypots one step further to emit signals proactively (as opposed to a traditionally very static and passive notion of honeypots which await actor inputs before emitting signals). Mirage theory attempts to use cognitive hacking [15] to influence the adversary's target selection process to force selection of incorrect targets and to highlight the attacker's presence in the network. Mirages also often add in an additional counter-deception protection mechanism by employing the digital/analog barrier to make it more difficult to detect the deception. As depicted in Fig. 4, there is typically a barrier where the analog signals that are transmitted by sensors and taken as input by actuators is converted to and from digital inputs/outputs. This conversion point is a horizon of sorts that disallows an adversary from directly perceiving

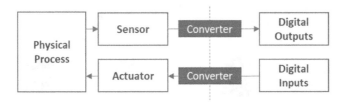

Fig. 4 Digital/analog barrier used in mirages

anything on the analog side of the system (left-hand side of Fig. 4). Thus, replacing the actual sensors, actuators, and physical processes with digital or analog simulators takes place behind a veil that cannot be easily pierced by an adversary. Mirage theory attempts to align with the principle from traditional warfare that deception should be integrated with operations, an attribute not typically true for honeypots [16, 17].

Deception can, of course, also be employed against an ICS for offensive purposes. Beyond the traditional phishing emails, watering hole attacks, and other common forms of offensive deception, ICSs are subject to several specific offensive deception techniques. Deceptive signals can be fed to physical devices, human-machine interfaces, or other components to cause systems or humans to take inappropriate and potentially damaging actions. Work has been done in describing these weaknesses (e.g., in state estimation systems' Bad Data Detection [BDD] components) and introducing countermeasures [18–21].

Advanced Concepts from Traditional Networks

Deception in traditional networks and IT infrastructure has recently begun to move beyond honeypots to more advanced techniques that push into various new areas as described in Fig. 3. One thrust of this new work has been in moving the application of deception from "next to" operational systems (as is the case with honeypots and mirages) to be in-line with operational systems (or giving the appearance of such). As seen in Fig. 5, honeypots and mirages tend to be deployed as additional systems or virtually as the illusion of additional systems. These systems may live near the real systems and help distract adversaries or may be deployed completely disconnected

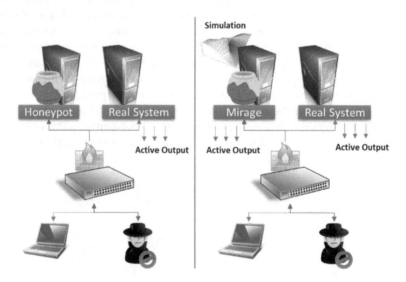

Fig. 5 Honeypots (left) and mirages (right)—operate as additional/parallel systems

and separate from any real system. This model has the nice benefit that as a visible, but non-real (from a business perspective) system, any interaction with them is reason for suspicion. Unfortunately, this also means that they are bypassable—a well-informed, sophisticated, or lucky adversary may never interact with them and may instead go directly to the real business systems. It is in this context that non-bypassable in-line deceptions have begun to emerge. Such a deception filter appears on the left in Fig. 6. Here, a transparent physical or virtual node inspects and potentially manipulates the traffic that flows through it. These filters inject deceptive content, timing, or other signals directly into the network stream. In this way, an adversary may attempt to interact with the real system, at its real network address, just as any legitimate client of software system would. A legitimate and malicious client may thus issue the same request, to the same host, but receive different responses. Such in-line deception filters have been implemented using software-defined networking [22] and as transparent proxies [23] among other possible forms. This model can be reactive/targeted (only applying deceptions to actors known to be malicious) or can be proactive (applying deceptions to all clients but (to be effective) applying deceptions that have outsized impacts on adversarial client interaction patterns). A variant of this technique, as depicted on the right in Fig. 6, can apply lightweight deceptions to all traffic in a proactive manner but, when a signal of malice is detected, can seamlessly transition select clients into an alternate reality that mirrors the original but is safely cordoned off from real systems. In this model, signals bearing network information (e.g., IP addresses) are manipulated to give the impression of continued interaction with the original node (the real system), though the traffic never touches that node after the detection event occurs.

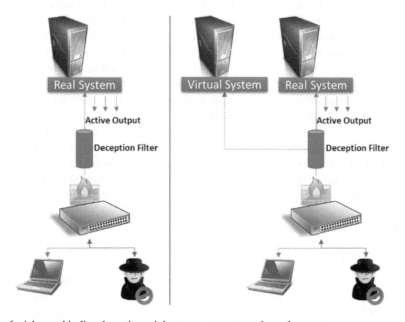

Fig. 6 Advanced in-line deceptions sit between, not next to, the real systems

Traditional network defense techniques, when successful, tend to remove unwanted intruders from the networks that they have infiltrated. An unfortunate side effect of such a strategy is to transition the defender-adversary relationship from a known state (i.e., adversary has been detected, source/target may be known, etc.) to an unknown state (i.e., all that is known is that the adversary is not where they were previously). In the case of well-resourced attackers, such as a nation state, the adversary is unlikely to end their efforts at such a setback but will far more likely simply reposition and retarget—attacking again, this time from an angle that the defender doesn't suspect. In the in-line alternate reality model described above, one can begin to address this deficiency of traditional defenses. Any detection event can move the adversary into the alternate reality where many of the well-accepted notions can be inverted. Here, one would like (1) an adversarial engagement to last as long as possible and (2) the adversary to use as many zero-day or other advanced attack techniques as possible. While somewhat counterintuitive, these properties stem from the fact that the alternate reality appears real yet is both safely sandboxed from real systems and closely monitored. This monitoring, when coupled with automatic remediation systems [24], can lead to a setup in which adversary actions undertaken in the alternate reality can automatically lead to patches and defenses applied to the real systems. Under such a system, the more energy (e.g., time, exploits, tactics) expended by an adversary, the stronger a defended system can become.

Some in-line parallel deception systems, such as presented in Soule et al. [22], go beyond adding a honeypot/mirage that clients can be seamlessly transitioned to (even after initial contact is made with a real system) to also embody the notion of active deception. In this context, one of the main goals is to keep the adversary engaged while in this alternate reality rather than off attacking real systems. Toward that end, rather than acting as a passive node, the reality can adapt to best prolong the adversarial engagement. As the instrumented system observes the interactions, it can automatically conform to the expectations and goals of the adversary. An SQL injection attack on a database, for example, can be met with deceptive maneuvers that cause the adversary to believe their attack worked but that they incorrectly guessed the wrong table name, leading to extended and fruitless activities (continued attack refinement), as discussed in Table 1.

Both the in-line and in-line parallel deception models exhibit the desirable property that the defense mechanisms can go beyond distraction to manipulate perception of the real system as a counter not only to reconnaissance but to many stages of various forms of attack [23]. In the subsections below, several classes of techniques that can be implemented via in-line or in-line parallel filters are described.

Table 1 Deception platform comparison

	Proximity to real system	Passive/active
Honeypot	Next to or completely separate	Passive
Mirage	Next to or completely separate	Active (fake data)
In-line filter	In front of	Active (fake and real data)
In-line parallel filter	In front of, as well as next to	Active (fake and real data)

Temporal Deceptions

Deceptions often involve the manipulation, creation, or hiding of content (i.e., hosts, ports, files, etc.). Deception can, however, also manipulate other properties of a signal. One such class of properties is temporal elements of a system/signal stream. In this case, the deception may manipulate latency information to make a system appear slower (or in some cases faster) than it really is or may manipulate the frequency of outputs. This type of deception can be useful for pre-attack anti-reconnaissance purposes (to disallow a true understanding of the temporal aspects of a system) but may also be used (often in conjunction with a stealthy, effective defense) to give a false impression of success to an attacker. If one is able to mount a successful defense against an attack, it may be beneficial to not let the attacker become aware of that. If an attacker believes that their attack is being effective, they are less likely to increase its strength or to attempt other, potentially unknown and undefended attack types. Consider, for example, a Denial of Service (DoS) attack attempting to achieve a specific level of degradation [25]. Such an attack may adapt its strength based on perceived attack efficacy. If through false latency manipulation one can give the impression that the system is failing, when in fact it isn't, the attacker may unwittingly take actions counter to their goals.

Temporal signals can take various forms. At the network level, measuring the round-trip time of an ICMP echo (ping) request/response is a common mechanism of latency measurement. In the case of a TCP-based request/reply interaction patterns, timing the client-perceived application level round trip (e.g., retrieval of a web page) can give another good indicator of network and system performance at the temporal level.

ICMP messages are small in size, and number, and thus in most cases, their latency can be easily managed simply by holding on to them (e.g., at a transparent in-line deception filter) until the desired delay has been reached. There are numerous implementation mechanisms to achieve this and various optimizations that can further reduce the deceiver's memory overhead in holding ICMP data for longer than would traditionally be required (e.g., only hold key information that cannot be otherwise derived at response time, such as source of the request and time of receipt).

At the TCP level, such hold and then release techniques can be costlier, as the amount of data to be held can grow to be very large, particularly in high-volume systems or during volumetric DoS attacks. To achieve the desired latency manipulation without requiring the deceptive node (e.g., an in-line deception filter) to maintain any state, an interesting use of the TCP retry mechanism can be employed. Other variants exist, but here we describe a latency induction technique called SYN dropping. The SYN drop maneuver allows a controlled disruption of the establishment of incoming TCP connections. Such a maneuver can be used to inject artificial latency (e.g., to give an adversary the false sense that their attack is successfully disrupting legitimate service) and can also be used to reduce adversary traffic volumes without denying their connections—a useful tactic that keeps the defense's awareness of the attack or attack source hidden. The TCP protocol undergoes a handshake during connection establishment. The initial steps of this handshake involve a client sending an SYN packet, followed by the server responding with an SYN-ACK, acknowledging receipt of the client's SYN. When an SYN packet is lost, or in this case intentionally dropped, the initiating client will not receive an SYN-ACK and thus, after a time-out period, will resend the SYN. Dropping SYN packets at a deception filter, for example, is one way to inject artificial latency without maintaining state on the server, as in this case, the client is responsible for keeping the packet content until it has been acknowledged as successfully received. The client's operating system is aware of the missed ACK, but at the application level, it simply appears to be a slower-than-normal connection establishment. This technique can be used to drop packets for specific IP addresses (targeted deception) or universally for all incoming requests. If a source IP is already under suspicion, then this maneuver can be narrowly targeted against that IP in a reactive manner. Traditional responses, such as blocking source IP addresses, are effective in limiting access for the given client but have the side effect of sending a signal to that client that their efforts are being mitigated. The advantage with the SYN drop maneuver, particularly in the context of cyber-deception, is that it can instead send a signal to the adversary that their attack is working and thus response time has been impacted. The SYN drop maneuver can be engaged proactively (i.e., not a targeted reaction to a detection of malice) as well. In this case, the impact will be driven by the interaction patterns of legitimate and adversarial clients. When such patterns differ (e.g., legitimate clients employ few, but long-lived connections while malicious clients employ many short connections), this mechanism can still be useful even without detection events [25].

Depending on the particular context, other sources of temporal information may exist, such as time stamps embedded in application-level messages. In such cases, given a sufficiently capable adversary, one may need to coordinate content-level deception with temporal-level deception to achieve a consistent output signal set.

Spatial Deceptions

Spatial deceptions, in the cyber-domain, manipulate the perceived sense of location, quantity, and status of network-reachable nodes. Such deceptions are used to mask the true structure of a network, to project false strengths or weaknesses, or to add sufficient noise to disallow the easy distinction between real assets and false assets. These false elements may manifest in a variety of ways. Real physical or virtual machines may be added to the network that are false only in the sense that they are not required to support business processing and exist as honeypots, for example, to (1) distract from the real business nodes, (2) to allow monitored interaction with adversaries, and (3) to be used for detection as interaction with such nodes is not expected from any legitimate client. Spatial deceptions may also exist purely in signal form, with no real manifestation as a physical or virtual node. In such cases, the perception of the node is projected from an in-network presence, such as the in-line deception filters discussed above. In this case, network traffic reaches a deception filter, which will respond on behalf of any number of real or false nodes. For example, an ICMP echo ping request for false hosts X, Y, and Z will all have a response

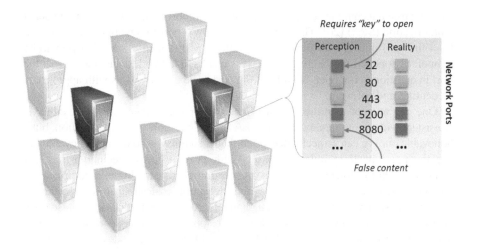

Fig. 7 Spatial deceptions—false hosts, false ports

generated in the filter, with the request propagating no further than the filter itself. Requests for real systems A and B may optionally also be handled in this filter. As depicted in Fig. 7, in addition to false hosts, other related deceptions can be projected, such as false open or closed network ports. A network port is a logical concept used by protocols such as TCP and UDP to specify a finer-grained endpoint than a host (as a single host may be running multiple network facing servers or clients). A port, which in most systems can take an identifier from 0 to 65,535, is only "open" if there is a network service that has taken action to listen at that specific port number. During the reconnaissance phase of an attack, an adversary often seeks to enumerate not only the available hosts in a network but the services that are running on each host. A common way to try to infer this service information is by inspecting which ports are open. A server with ports 22 and 80 open is most likely running an SSH server and a web server. Tools such as nmap and zmap are often used to enumerate such information. Further refinement of this information can be undertaken by attempting initial interaction with each port and examining the resulting response, for example, sending an HTTP GET request and seeing indicators that an Apache HTTPD server is running or getting to the log in prompt for an SSH server and seeing a welcome banner describing the system as being Ubuntu version 16.04.

As a result of these techniques, an adversary can relatively easily map out accessible network nodes, and the operating systems, and services running on each. Projecting false hosts can lead to incorrect network topology inference, and projecting false open ports and false initial content on those ports can lead to incorrect host operating system and service inference. False ports can be projected both on real nodes, altering the perception of an actual asset, and on false nodes, leading to a more realistic false node projection. False nodes and ports can also be good indicators of reconnaissance activities as legitimate clients should have no reason to interact with them. Ports that are in fact "open" can also be made to appear closed through techniques based on the concepts of port knocking and Single Packet Authorization (SPA) [26]. These maneuvers allow a port to remain closed (blocked by a firewall) until a user presents a specific key that then allows that port to be temporarily opened for the given user. This results in network scans finding that the port in question is not open but allows the port to be used by users with appropriate credentials and software.

One can attempt to project a false reality over these dimensions that either is consistent with common network and system configurations (e.g., a deception, but a believable one) or can project a deception that is so noisy as to not be believable but use the noise itself to hide real assets within. An effective technique has, for example, been to project all ports as open, significantly slowing and complicating reconnaissance efforts. When a believable deception is desired, it is important to realize consistency across observation channels. An adversary that saw a false open port on host X, but saw that host X does not respond to ping requests, may use that discrepancy as a tell, indicating that host X is likely a false node. As depicted in Fig. 8, work has been done as described by Pal et al. [24] in allowing the output of network design tools to act as input to in-line deception filters, which can then project a consistent reality across all channels.

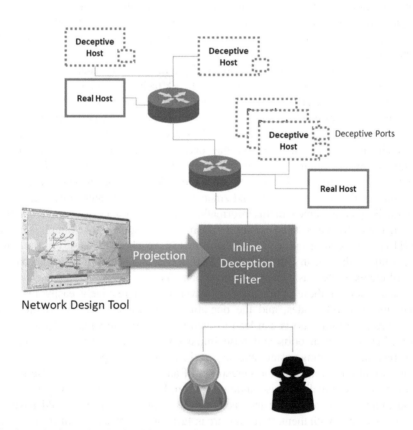

Fig. 8 Designing and projecting false network structures

Client Validation and Manipulation

Deception can be employed in a subtle, stealthy manner to validate the nature of a client or to manipulate a client such that clients of one nature see a negligible effect while clients of another nature see a more pronounced effect. This suite of techniques can be used to help direct other actions through achieving more confidence that a client is legitimate or malicious (in the case of validation) or to effectively throttle attacks (in the case of manipulation).

Client validation can be used as a double check to ensure the client that one is interacting with fits the profile and characteristics of an expected client. For example, as shown in Fig. 9, in the domain of web browsing, a sophisticated full-featured and modern browser likely exhibits different behaviors from the attack software being employed by a botnet. When a browser requests a web page, it will first request HTML and then soon after request CSS, JavaScript, images, or other artifacts that should be embedded within the page content. Further, when the JavaScript arrives at the browser (if scripting is not blocked), the browser will potentially execute some amount of JavaScript code. A malicious client attempting, for example, an HTTP flood attack would, on the other hand, execute a series of more frequent requests,

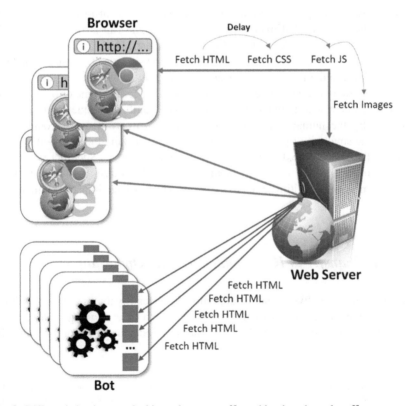

Fig. 9 Differentiating between legitimate browser traffic and bot-based attack traffic

often for the same exact resource (e.g., just the HTML). Since rendering the returned content is not of interest to the attacker, the client software is likely not to be a full browser, and the HTML content is unlikely to be interpreted, or even parsed, and thus embedded resources not fetched or executed. Even at the HTTP level, the response codes (that might indicate an error or a redirection) may not be inspected or acted upon. All of these differences are ripe for analysis in order to determine if the client appears to match the behavior of a legitimate client or not. Deception can be employed here to inject specific canary resources that if not fetched, executed, or otherwise acted upon would indicate a client that does not match legitimate profiles. An in-line deception filter could manipulate the HTML response to a request to inject a reference to a small image, CSS, or JavaScript file and could watch for subsequent requests of the marker resource. If such a subsequent request is not observed, confidence in the client's legitimacy could be diminished, and other deceptive or nondeceptive defenses could be employed as a result. Other similar mechanisms can be used that will set a cookie that is checked for on subsequent requests from the client (ensuring the client software at least interprets and handles cookies) or set a 300-level return code indicating that the resource requested is at a new location and watch for a subsequent request at that location to ensure the client software interprets and follows HTTP redirects.

The above techniques manipulate the response to a client to help validate the nature of the software that made the request. Deception can also be used to help minimize volumetric attacks that employ many requests. Here, an in-line deception filter can be used, in the example domain of web browsing again, to inject small bits of JavaScript into the response to a web request. This JavaScript can then execute within the browser and consume a small amount of processing time or memory. For a legitimate client who is making one, or a small handful of requests, this will be unnoticed by the human user. For a client making many frequent connections, however, the small bit of processing will add up to a meaningful level—effectively throttling the attack by exhausting or taxing client resources. Such deceptions are meant to make the cost to the client higher for each request, at a level that has minimal impact on legitimate usage patterns yet outsized impact on malicious patterns.

Potential Transfer to ICS, Challenges, and Opportunities

The malleability of software and the ubiquitous connectivity that is available today is a potent combination that has led to the phenomena of everything as a service,[1] where functionality and features that were once realized in hardware or mechanically

---

[1]Recovery (or disaster recovery) as a service (RaaS or DRaaS), a recent newcomer in the marketplace compared to IaaS or PaaS, is said be worth billions of dollars of business in June 2017 (https://en.wikipedia.org/wiki/Recovery_as_a_service).

are now implemented in software that can be materially removed from the physical elements or the users of the system. This is both a bliss and a curse. On the positive side, it makes integration of different types of devices, plants, and systems easier; controllers and data aggregators can be distributed; data can be collected faster and in more detail, thereby enabling better and more rapid responses. And, most relevant for this chapter, everything as a service makes it easier and cheaper to stand up deceptive mechanisms like mirages. On the negative side, the ease of connecting things and the drive to market quickly increase the proclivity to repeat the mistakes made in mainstream IT systems such as going live, without a full understanding of the implication or improper and poor implementation of air gaps. Since more systems are made to interoperate and interact, many parts of the system that were previously not remotely accessible are becoming so. It is in this background and with the acknowledgment that mirages are a well-studied construct in the ICS context we begin our discussion about adopting and adapting deceptive maneuvers from traditional networked systems to ICS networks.

Let us consider a hypothetical ICS for a power company that among other elements operates a wind farm as shown in Fig. 10. Each generation station, including the wind farm, has its own local area network with an on-site consolidator and router node. Multiple such nodes feed into the control station over a wide area network (WAN), which, for this example, we will assume is a proprietary network, but it will not be farfetched to think of that WAN procured from a network vendor as a NaaS (network as a service) that is tunneling the power company's traffic over a shared network infrastructure using some form of VPN (or optical, if the lines are fiber optics) technology to isolate it from other co-tenants. The control center hosts the company's HCI systems as well as enterprise data storage and servers, which are supposed to be air-gapped (as shown by the one-way arrow in the figure) from the HCI (and the rest of the plant-side elements of the ICS). The enterprise server and data storage are available through the Internet to the company's corporate data and control centers for analytics and overall management. Overlaid in this diagram are a few mirage wind turbines that exist only virtually—think of simulation or high-fidelity models of wind turbines that exist only in software receiving control input and producing observable output except for the actual current that is fed to the step-

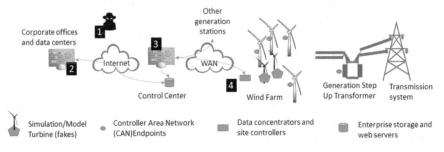

Fig. 10 ICS of hypothetical power company

up transformer at the wind farm. For simplicity, let us assume that the computers hosting the simulation models of the turbines are physically located in the wind farm.

Also overlaid in this diagram is a simplified threat model, indicated by the attacker icon and the numbers with a black background to help us understand where deceptive maneuvers are applicable. For each of these situations, we ask: (a) what harm can the attacker cause to the ICS, and (b) what deceptive maneuvers can we use to defend against such harm? Here, we consider four main scenarios, each based off of different starting point assumptions:

1. The attacker is in the Internet and has no credential or access to the power company.
2. The attacker has compromised a corporate user's computing node via social engineering, waterhole attacks, or other means.
3. The attacker has breached the control center, where there are two sub-scenarios:

   (a) He is on the Internet side of the air gap.
   (b) He is on the plant side of the air gap.

4. The attacker has progressed to the wind farm (i.e., he is on the local Modbus, Profibus, or similar network) and has breached a system like the site controller on the premises of the wind farm.

In answering the two questions we posed earlier for each of the four scenarios, we will highlight the differences from traditional systems—looking at what, if anything, becomes easier or more difficult in the ICS network.

Attacker in the Internet

Without any specific privilege or access to the systems at the corporate offices or the control center, the adversary will have a few options:

(a) Mapping Internet-accessible systems in the corporate offices or at the control center
(b) Identifying legitimate users with access to corporate systems and applying targeted social engineering or other attacks against them to gain access at their level
(c) Mounting attacks to gain a toehold in identified vulnerable systems that are accessible from the Internet
(d) Mounting attacks (i.e., DDoS, Ransomware, etc.) against the power company's enterprise systems to disrupt its regular operation

Step (a) or (b) is usually a precursor to (c) or (d). And although it is hard for the adversary to directly damage the plant or physical devices (e.g., wind turbines or the step-up generator), careful readers will recognize that step (c) gets the attacker one step closer to achieving that goal and that step (d) can be used to cause indirect damage if the power company does not have the proper procedures—for example, the plant-side ICS failing to update the enterprise data storage may result in cascading failure further upstream, or the corporate systems being unable to access the enterprise data and services at the control center may result in incorrect energy trading decisions which, in addition to financial issues, may stress the grid.

Deceptive maneuvers that are applicable in mainstream IT systems and networks can easily be applied at this stage. The interaction between corporate users and the enterprise system at the control center is most likely to be web-based, riding over HTTP or HTTPS. Apart from honeypots in the corporate and the control center networks, or fake user accounts, the following deceptive maneuvers are also possible:

- *In-line parallel deception filters*: Create an alternate reality that hides the real systems from untrusted users yet allows malicious actors to execute in a realistic-looking context where they can be observed and those observations used to strengthen defenses. Under our Keeping the Adversary Guessing and Engaged (KAGE) work [22], we realized a construct using virtualization and software-defined networking (SDN) for doing so. In this construct, a service endpoint (at IP address x, port y) different than the real service endpoint (at IP address i and port j) is advertised to everybody as the provider of the protected service, service requests from selected (based on static white list or dynamically derived based on behavior) corporate IP addresses directed to (x:y) are sent to (i:j), and responses from (i:j) are made to appear to have come from (x:y) (as shown in Fig. 11). In the context of an ICS, one typically finds a far more restricted set of allowed users than in traditional public-facing services and networks. This property allows the application of white listing in more use cases and thus supports such schemes with less risk of false positives and negatives. In addition, the protected service at (i:j), running inside a virtual machine container, is monitored using virtual machine introspection (VMI) and a strict application-specific policy that can kill the VM and restart the service from a checkpointed state as soon as a policy violation is observed. Any traffic that is not a service request addressed to (x:y) is delivered to (x:y). As a result, attacks from IP addresses outside of the white list never reach the real services. At the same time, attacks that appear to come from the white-listed corporate IP addresses

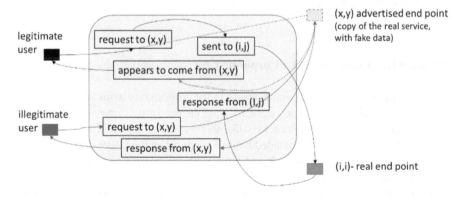

Fig. 11 ICS of hypothetical power company

(either through spoofing or compromised corporate systems) also have a hard time corrupting the target service or data.

- *Expanding the attacker's search space*: Present a completely different network full of open ports and potentially vulnerable services to the adversary's attempt to map the corporate network or the services at the control center without having to expend computing resources that are needed to create honeypots. Our prior work in Adaptive Resource Management Enabling Deception (ARMED) [24] demonstrated how this can be achieved using transparent network proxies for ICMP, TCP, and UDP.
- *False disruption indicators*: By introducing specialized network actors in the defended network enclave, it is possible to detect when specific services are victims of DoS by sensing request rates as well as stresses on system resources. These protocol-specific network actors, as we demonstrated in our ARMED work, introduce a control point for a given network protocol to monitor, check, and adapt the protocol behavior transparently (i.e., the endpoints interacting over the subject protocol are not aware of the network agent intermediary). In response to this detected stress, the network actors can be used to perform maneuvers like probabilistic SYN DROPs of varying intensity and targets (source IPs in a TCP SYN request), which cause controllable delays in TCP interactions without breaking the legitimate clients' expectation but giving the adversary clients the false impressions that the victim systems they are trying to connect to are getting bogged down.
- *Injecting additional work to uncover bots*: The protocol-specific network actors can also be used proactively to inject small amount of redirection or additional work on the clients requesting services. For example, as we showed in our ARMED work, network actors specialized for the HTTP protocol can be used to inject a JavaScript URL redirection with a cookie in response to a HTTP request. If the request was from an attack bot, chances are that the bot does not implement the full browser functionality and will not follow the redirection and present the cookie in its subsequent attack requests, which can then be dropped. This is one way to use deceptive maneuvers to defend HTTP-based services in the power company's corporate network.

Attacker Has Compromised a Corporate User

At this stage, the attacker has some access to the corporate systems. Depending on the user he was able to compromise, he may use the enterprise systems at the control center directly. Sophisticated attackers like to stay hidden and find indirections or additional ways to attack the intended target T, even when they have direct access to T. The attacker may try to breach additional systems in the corporate network or expand his access into the enterprise.

In addition to using an in-line parallel deception filter to completely hide a real server endpoint at IP address i and port j, and advertise the service at IP address x and

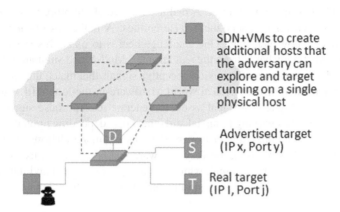

SDN+VMs to create
additional hosts that
the adversary can
explore and target
running on a single
physical host

Advertised target
(IP x, Port y)

Real target
(IP I, Port j)

Fig. 12 Expanding the exploration space for the adversary

port y as explained earlier, it is possible to create intermediate hosts that are reachable from the corporate network that can reach IP address x but not i. This will incur computing cost since we would want the attacker to be able to log in to the intermediate host—but using software-defined networking and virtualization, it is possible to create a number of such hosts on a single physical host organized in a complex network topology with indirect access to the host at IP address x (see Fig. 12). Note only service requests from specific corporate source IPs are sent to the real endpoint (IP address i, port j). So any attempt to touch the advertised service endpoint (IP address x, port y) from these additional hosts will always go to the fake endpoint (IP address x, port y) and not the real endpoint (IP address i, port j). This is one way to use deception to increase the exploration space for the attacker and keep him engaged within these intermediary hosts while the real target T remains behind the smoke screen.

If the adversary is able to compromise a user U that can access the enterprise systems hosted at the control center from the corporate network over the Internet, and if the adversary can invoke operations that U is entitled to, including read and possibly write (or other side effect-generating) operations on the enterprise database and services, then white list-based deceptive maneuvers will not be applicable. Here, the connecting client is in fact in the white list but is being controlled by adversarial actors. In this case, deceptive maneuvers are still feasible under the following situations: (1) deception based on a trigger, such as abnormal behavior or data patterns, or (2) deception that does not have negative side effects on legitimate clients. The first case would rely on existing detection mechanisms, or other deceptions (e.g., honeypots, honey-files, honey-ports, honey-web-pages, etc.), to enable deception for a particular interaction or client. There is risk here of false positives and unintentionally deceiving legitimate clients, however, so great care must be taken and verification or double checks may be warranted. In the second case, deceptions may be applied that have a detrimental effect only for malicious clients. For example, one may present ICS-specific misinformation such as the number and status of turbines in the wind farm. A real user, aware of the properties of the plant, would not search for or query these false entities. A malicious client performing reconnaissance, however, may. One challenge here is to make this misinformation credible—indistinguishable from the real turbines. Models and simulation have been used in such circumstances to provide deceptive signals that match the constraints of the real systems and of the physical world. Hardware-in-the-loop (HiL) can also be used here to employ real physical assets to generate realistic false signals, though such an approach has varying financial and other costs depending upon the hardware in question. Building false signals as perturbations and manipulations of real signals can help address this by using the very hardware that is actually in use for business purposes to generate a base signal, which can then be duplicated and then modified in hardware or software to create multiple derivative deceptive signals.

Attacker in the Control Center

The key cyber-physical elements of an ICS may be separated by an air gap. However, true and perfect air gaps are rare, many have unintentional/unknown breaches, and above all, there is a human element—the human operators can knowingly or unknowingly bridge the air gap by plugging a laptop in on both networks or copying files from one to another.

Even without bridging the air gap, the adversary on the Internet side can create a problem on the plant side (as depicted in Fig. 13). By creating load on the database or server, the adversary can block the plant-side updates resulting in dropped messages or even exceptions upstream if the plant-side software is not programmed robustly. Depending on how the plant-side software accesses the server, it may also be possible to mount a watering hole-style attack where the plant-side program may catch a malicious script. While there are many engineering solutions to counter these attacks, in terms of deception, we would like to highlight the possibility of using SDN to create the impression of the topology shown in Fig. 13, whereas the real topology is like what is shown in Fig. 14.

If the attacker compromised a user account and the user accessed corporate systems from the plant side (knowingly or unknowingly, e.g., to access active directory or a software update server, as shown in Fig. 15), then the attacker can

Fig. 13 Adversary on the Internet side of the air gap may impact the plant side

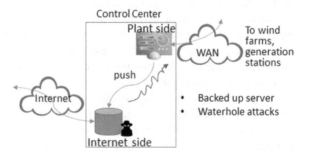

Fig. 14 Creating alternate reality for the air gap

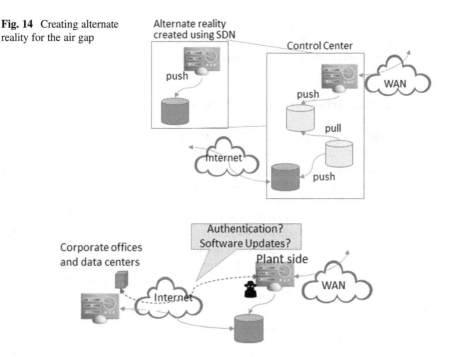

Fig. 15 Adversary may gain access to the HCI systems on the plant side of the air gap

control the user's keyboard/mouse and observe what he/she is doing or initiate malicious activities with his/her credentials.

Since the attacker is not in the control center physically, it is possible to create a mapping table that is accessible to legitimate users sitting at the control center and not the attacker who may be controlling the keyboard or mouse of the HCI console. The mapping table would act as a key and may identify, for example, which turbines are real and which are fake and whether the power outputs attributed to the fake ones are real (e.g., a fraction of the output of a real turbine can be attributed to a fake one, with that amount subtracted from the real turbine) or not. With support from mapping tables like this, it is possible to execute deeper deception, where simulated turbines can appear as real turbines to the remote adversary with keyboard and mouse control of the HCI workstations. Such deception keys can allow legitimate and malicious users to look at the same signal/information yet only deceive malicious actors. Similar principles have been applied in deceptive file systems, where the real file systems are hidden among a false one and without the key, one cannot distinguish between the two (or in some cases even see the true entries at all).

We note that the mapping table can be used in computation as well, not just as a means for human operators to look things up. However, it needs to be treated as an out-of-band mechanism. For example, if the data about the simulated turbines are pushed to the enterprise databases, users and analytics that need the actual number or performance data of turbines for their job functions must also use the mapping tables to retrieve the real information. The out-of-band requirement is needed to ensure that the presence and use of the mapping tables are not plainly visible to the adversary who is progressively gaining access into the corporate and ICS networks.

Attacker on the Wind Farm's Local Network

If the attacker has progressed from the plant side of the control center air gap to the local SCADA or controller area network at the wind farm, he has likely landed on a router or the data concentrator box. With that access, he may try to escalate his privilege to act as the site controller, map the elements of the network, drop specific messages, or interact with network elements using protocol-specific messages. Even though at this point he has a lot of power to disrupt the operation of the turbines or the transformer, this is the context where using mirage (i.e., creating mirage turbines or transformers) is most appropriate. In our description so far, we have used high-fidelity simulation models to create the mirage turbines. If models of transformers exist, they can also be used. One of the challenges in this approach is that the simulations run on (relatively powerful) general-purpose computers (as opposed to other approaches that use, e.g., hardware-in-the-loop simulators), and they must not be visible as compute servers on the local network because otherwise, the mirage or deception will be broken. In some scenarios, mirages hide behind the digital/analog boundary (as described above) to hide such resources. When that approach is not possible, we can rely on SDN and network virtualization again; since the attacker is not on the wind farm physically, it is possible to hide the computer servers from scans and probes or make them appear like a Modbus TCP endpoint or an Ethernet device that looks like other wind turbines.

Conclusion

We conclude this chapter with a number of observations. First, the boundary where the corporate network ends and the ICS network begins is beginning to get blurred and often consists of a narrow air gap that human operators can easily bridge either intentionally or unintentionally. One should not rely on the air gap. One should also not rely solely on deception to address the threat of cyberattacks on ICS, but deception, as outlined in this chapter, should be considered in designing the security protection at all strata of an organization's network. Second, compared to traditional and general-purpose networks, ICSs make use of a lot of white listing (e.g., which IP addresses can use specific services), particularly as you get closer to the physical devices, which, as we showed in the previous discussions, helps deceptive maneuvers. Third, most ICS protocols do not employ encryption or cryptographic authentication in the network segments closest to the devices/control. This is a mixed blessing because it makes the ICS more vulnerable to snooping or data manipulation, but it also makes it easier to perform deceptive manipulation of real traffic which needs access to the unencrypted content to be able to use it as a base, to be manipulated, or to understand the request such that a deceptive response can be composed. Fourth, there are a great variety of potential protocols in use (e.g., Common Industrial Protocol [CIP], Modbus, DNP3, Profibus, EtherCAT, etc.). From the deceptive maneuvering point of view, we hypothesize that this presents an opportunity to create alternate realities that could present, for example, an actual Modbus system as a Profibus system, although work remains to experimentally prove such an approach. Fifth, mirages seem to be the de facto approach to deception near the devices—although the purpose and means vary. We claim that when combined with SDN/virtualization, high-fidelity simulation has a lot of potential to create deep deception that can be very hard to disambiguate by a remote adversary. Sixth, concepts like in-line deception filters to create alternate realities, injecting controlled probabilistic failures that are hard to disambiguate from normal behavior (such as a packet dropping) or injecting additional work within protocol specification (such as an HTTP redirect or TCP retransmission), often destabilize the attacker's automated scripts and programs. Finally, even though deceptive maneuvers exhibit a lot of potential benefits in cyber-defense, there are a number of challenges that require further R&D. One of these challenges is the CAPEX (what additional investments are needed to use deception; covering hardware, software, and man-power) and OPEX (does deception slows down legitimate use of the system) over-heads and how to minimize them. A better question in this context may be how we determine the cost-benefit trade-off or the design and operational "sweet spot" for using deception. This leads to another critical question about deception that is only now being truly explored: How do we measure and evaluate the utility and effec-tiveness of defensive cyber-deception? We have developed a white-boarding-style approach to assess the impact of deception on a multistage attack workflow, which, like any white-boarding analysis, is a manual- and expertise-intensive approach and only offers a quantitative view of the expansion of the attacker's decision space.

References

1. I. Colvin, The Unknown Courier: The True Story of Operation Mincemeat, London, England (Biteback Publishing, 2016)
2. D. J. Bacon, *Second World War Deception. Lessons Learned for Today's Joint Planner*, (Air Command and Staff Coll Maxwell AFB, AL, 1998)
3. F. Cohen, D. Lambert, C. Preston, N. Berry, C. Stewart, E. Thomas, *A Framework for Deception – Final Report*, (2001)
4. B. Whaley, *Detecting Deception: A Bibliography of Counter-Deception Across Time, Cultures, and Disciplines*, (2005)
5. G. Stein, *Encyclopedia of Hoaxes* (Gale Research, Inc., Detroit, 1993), p. 293
6. B. Cheswick, An evening with Berferd – In which a cracker is lured, endured, and studied, in *USENIX Winter Conference*, (1992)
7. S. M. Wade, SCADA Honeynets: The attractiveness of honeypots as critical infrastructure security tools for the detection and analysis of advanced threats, Master's Thesis, (Iowa State University, Ames, IA, 2011), http://lib.dr.iastate.edu/cgi/viewcontent.cgi?article=3130& context=etd

8. V. Pothamsetty, M. Franz, SCADA Honeynet Project: Building honeypots for industrial networks, (SCADA Honeynet Project, 15 July 2005), http://scadahoneynet.sourceforge.net/
9. T. Vollmer, M. Manic, Cyber-physical system security with deceptive virtual hosts for industrial control networks. IEEE Trans. Industr. Inform. **10**(2), 1337–1347 (2014)
10. Digital Bond, Installation instructions virtual PLC Honeynet, (2006), http://www.digitalbond.com/blog/2011/07/27/siemens-s7-honeynet/#more-10410
11. N.C. Rowe, J. Rrushi, *Introduction to Cyberdeception* (Springer International Publishing, Cham, 2016)
12. Attivo Networks, The benefits of deception for SCADA environments, (2018), https://attivonetworks.com/benefits-deception-scada-environments/
13. Illusive Networks, Turn each endpoint into an APT trap, (2018), https://www.illusivenetworks.com/deceptions-everywhere/
14. J. Rrushi, K. Kang, Mirage theory: A deception approach to intrusion detection in process control networks, in *Proceedings of the NATO Symposium on Information Assurance for Emerging and Future Military Systems*, (2008)
15. J.L. Rrushi, An exploration of defensive deception in industrial communication networks. Int. J. Crit. Infrastruct. Prot. **4**(2), 66–75 (2011)
16. N.C. Rowe, H. Rothstein, *Deception for Defense of Information Systems: Analogies from Conventional Warfare*, (Department of Computer Science and Defense Analysis, U.S. Naval Postgraduate School, 2003), http://www.au.af.mil/au/awc/awcgate/nps/mildec.htm
17. N.C. Rowe, H. Rothstein, Two taxonomies of deception for attacks on information systems. J. Inf. Warf. **3**(2), 27–39 (2004)
18. A. Teixeira, G. Dán, H. Sandberg, K.H. Johansson, A cyber security study of a SCADA energy management system: Stealthy deception attacks on the state estimator. IFAC Proc. **44**(1), 11271–11277 (2011)
19. S. Amin, X. Litrico, S. Sastry, A.M. Bayen, Cyber security of water SCADA systems—Part I: Analysis and experimentation of stealthy deception attacks. IEEE Trans. Contr. Syst. Technol. **21**(5), 1963–1970 (2013)
20. S. Amin, X. Litrico, S. S. Sastry, A. M. Bayen, Stealthy deception attacks on water SCADA systems, in *Proceedings of the 13th ACM International Conference on Hybrid Systems: Computation and Control*, (2010), pp. 161–170
21. A. Kleinmann, O. Amichay, A. Wool, D. Tenenbaum, O. Bar, L. Lev, Stealthy deception attacks against SCADA systems, in *Third Workshop on the Security of Industrial Control Systems & Cyber-Physical Systems (CyberICPS)*, (Oslo, Norway, LNCS 10683, 2017), pp. 93–109
22. N. Soule, P. Pal, S. Clark, B. Krisler, A. Macera, Enabling defensive deception in distributed system environments, in *IEEE Resilience Week (RWS)*, (2016), pp. 73–76
23. P. Pal, N. Soule, N. Lageman, S. S. Clark, M. Carvalho, A. Granados, A. Alves, Adaptive resource management enabling deception (ARMED), in *Proceedings of the 12th International Conference on Availability, Reliability and Security*, (2017), p. 52
24. P. Pal, R. Schantz, A. Paulos, B. Benyo, D. Johnson, M. Hibler, E. Eide, A3: An environment for self-adaptive diagnosis and immunization of novel attacks, in *2012 IEEE Sixth International Conference on Self-Adaptive and Self-Organizing Systems Workshops (SASOW)*, (2012), pp. 15–22
25. P. Pal, N. Lageman, N. Soule, Disrupting adversary decision logic: An experience report, in *ECCWS2018-Proceedings for the 17th European Conference on Cyber Warfare and Security*, (Academic Conferences and Publishing Limited, 2018)
26. M. Rash, Single packet authorization with fwknop.login. USENIX Magazine **31**(1), 63–69 (2006)

Moving Target Defense to Improve Industrial Control System Resiliency

Adrian R. Chavez

Abstract Historically, control systems have primarily depended upon their isolation from the Internet and from traditional information technology (IT) networks as a means of maintaining secure operation in the face of potential remote attacks over computer networks. However, these networks are incrementally being upgraded and are becoming more interconnected with external networks so they can be effectively managed and configured remotely. Examples of control systems include the electrical power grid, smart grid networks, microgrid networks, oil and natural gas refineries, water pipelines, and nuclear power plants. Given that these systems are becoming increasingly connected, computer security is an essential requirement as compromises can result in consequences that translate into physical actions and significant economic impacts that threaten public health and safety. Moreover, because the potential consequences are so great and these systems are remotely accessible due to increased interconnectivity, they become attractive targets for adversaries to exploit via computer networks. Several examples of attacks on such systems that have received a significant amount of attention include the Stuxnet attack, the US-Canadian blackout of 2003, the Ukraine blackout in 2015, and attacks that target control system data itself. Improving the cybersecurity of electrical power grids is the focus of our research.

Introduction

Historically, control systems have primarily depended upon their isolation [1] from the Internet and from traditional information technology (IT) networks as a means of maintaining secure operation in the face of potential remote attacks over computer networks. However, these networks are incrementally being upgraded [2] and are becoming more interconnected with external networks so they can be effectively managed and configured remotely. Examples of control systems include the

A. R. Chavez (✉)
Sandia National Laboratories, Albuquerque, NM, USA
e-mail: adrchav@sandia.gov

© Springer Nature Switzerland AG 2019
C. Rieger et al. (eds.), *Industrial Control Systems Security and Resiliency*, Advances in Information Security 75, https://doi.org/10.1007/978-3-030-18214-4_8

electrical power grid, smart grid networks, microgrid networks, oil and natural gas refineries, water pipelines, and nuclear power plants. Given that these systems are becoming increasingly connected, computer security is an essential requirement as compromises can result in consequences that translate into physical actions [3] and significant economic impacts [4] that threaten public health and safety [5]. Moreover, because the potential consequences are so great and these systems are remotely accessible due to increased interconnectivity, they become attractive targets for adversaries to exploit via computer networks. Several examples of attacks on such systems that have received a significant amount of attention include the Stuxnet attack [6], the US-Canadian blackout of 2003 [7], the Ukraine blackout in 2015 [8], and attacks that target control system data itself [9]. Improving the cybersecurity of electrical power grids is the focus of our research.

The power grid is responsible for providing electricity to society, including homes, businesses, and a variety of mission-critical systems, such as hospitals, power plants, oil and gas refineries, water pipelines, financial systems, and government institutions. The "smart grid" acts as an advanced power grid with upgrades that provide power distribution systems and consumers with improved reliability, efficiency, and resiliency [10]. Some of the upgrades include automated energy load balancing, real-time energy usage tracking and control, real-time monitoring of grid-wide power conditions, distributed energy resources, advanced end devices with two-way communications, and improved processing capabilities. Advanced end devices, which are being integrated into smart grids, include programmable logic controllers (PLCs), remote telemetry units (RTUs), intelligent electronic devices (IEDs), and smart meters that are capable of controlling and performing physical actions, such as opening and closing valves, monitoring remote real-time energy loads, monitoring local events such as voltage readings, and providing two-way communications for monitoring and billing, respectively. These new devices replace legacy devices that have been in place for decades that were not originally designed with security in mind since they were previously closed systems without external network connectivity. Although these new devices aid in efficiency, they may create more avenues for attack from external sources.

Additionally, control systems are often statically configured [11] over long periods of time and have predictable communication patterns [12]. After installation, control systems are often not replaced for decades. The static nature combined with remote accessibility of these systems creates an environment in which an adversary is well positioned to plan, craft, test, and launch new attacks. Given that the power grid is actively being developed and advanced, the opportunity to incorporate novel security protections directly into the design phase of these systems is available and necessary. Of particular interest are defenses that can better avoid both damage and loss of availability, as previously documented in the power grid [5], to create a more resilient system during a remote attack over computer networks.

Challenges

One of the main challenges of introducing modern computer security protections into industrial control systems (ICSs) is to ensure that the computer security protections themselves not only improve the security of the overall system but also do not impede the operational system from functioning as expected. A security solution that is usable and practical within an IT environment may not necessarily be practical within an ICS environment. ICSs often have real-time requirements, and any newly introduced software or security solution must also meet those same requirements.

Another challenge is to identify useful metrics that quantify the effectiveness of the moving target defense (MTD) techniques from the perspective of both the adversary and the defender of the system. The goal of the adversary is to exploit the system before the MTD defends against the attack by modifying the environment. The goal of the defender is to change the environment frequently enough to evade an adversary, but not too frequently so that system performance is negatively impacted. Finding the correct balance is necessary so that the adversary cannot exploit the system and the MTD strategy does not prevent the system from maintaining a normal operating state.

Gaining access to representative ICS environments is another challenge when developing new security protections for ICS systems. Modeling and simulation tools can be effective, but gaining a true understanding of the consequences and effects of deploying a new security protection in practice requires validation within a representative ICS environment. Several factors such as network load, processor load, and memory load are difficult to accurately project within a simulated environment. The harsh working conditions of ICSs (such as wide temperature ranges) are one element to consider when deploying new technologies within these environments.

MTD Within Critical Infrastructure

Critical infrastructure systems bring in a distinctive set of constraints and requirements when compared against traditional IT-based systems. Critical infrastructure systems are often time-sensitive with stringent real-time constraints, as is the case for cyber-physical systems [13]. It is therefore important for any new computer security protections introduced to also meet these same time requirements so that they do not negatively affect the operational network. Additionally, the most important requirements for these systems are often to maintain high availability and integrity due to the nature of the systems that they control (e.g., the electrical power grid, water pipelines, oil and natural gas refineries, hospitals, residential and commercial buildings, etc.). Any loss of availability can result in significant consequences not only in terms of economics but also in terms of public health and safety. Similarly, compromising the integrity of these systems, such as sending maliciously modified commands, can result in similar consequences. Also of note is that critical infrastructure systems are composed of both legacy and modern systems that must interoperate with one another without affecting availability and security. New security solutions must therefore take interoperability into account so that they can scale without the requirement of upgrading every end device within the system. Additionally, in order for MTD-based approaches to be successfully deployed within critical infrastructure environments, they must satisfy the distinct time constraints and requirements of these environments. Since the time constraints vary from one system to another, the stricter time requirement used for teleprotection systems should be used (12–20 ms) [14]. For Supervisory Control and Data Acquisition (SCADA) communications, those requirements can, in some cases, be relaxed to 2–15 s or more.

Background

Artificial diversity is an active area of research with the goal of defending computer systems from remote attacks over computer networks. Artificial diversity within computer systems was initially inspired by the ability of the human body's natural immune system [15] to defend against viruses through diversity [13]. Introducing artificial diversity into the Internet Protocol (IP) layer has been demonstrated to work within a software-defined network (SDN) environment [16]. Flows, based on incoming port, outgoing port, incoming media access control (MAC), and outgoing MAC, are introduced into software-defined switches from a controller system. The flows contain matching rules for each packet and are specified within the flow parameters. If a match is made within a packet, then the flow action is to rewrite source and destination IP addresses to random values. The packets are rewritten dynamically while they are in flight, traversing each of the software-defined switches. Although applying artificial diversity on an SDN has been demonstrated, the effectiveness of such approaches has not been quantitatively measured. Furthermore, to our knowledge, the approach has not been deployed within an ICS setting, which differs substantially from traditional IT-based systems.

It has also been demonstrated that IP randomization can be implemented through the Dynamic Host Configuration Protocol (DHCP) service that is responsible for automatically assigning IP addresses to hosts within the network [17]. Minor configuration modifications to the DHCP service can be made to specify the duration of each host's IP lease expiration time to effectively change IP addresses at user-defined randomization intervals. However, this approach only considers long-lived Transmission Control Protocol (TCP) connections; otherwise, disruptions in service will occur as the TCP connection will need to be re-established. Service interruptions within an ICS setting is not an option due to their high-availability requirements. Quantifying the effectiveness of such approaches has also not been performed within an ICS setting outside of surveys [18] that evaluate MTD techniques within an IT setting where IP randomization by itself was qualitatively ranked to have low effectiveness with low-operational costs. Also of note is that IP randomization approaches by themselves have been demonstrated to be defeated through traffic analysis where endpoints of the communication stream can be learned by a passive adversary, observing and correlating traffic to individual endpoints [19].

Anonymization of network traffic is an active area of research with several implementations available in both commercial and open-source communities. MTD and anonymization are related in that they both have the goal of protecting attributes of a system from being discovered or understood. One of the early pioneering groups of anonymous communications describes the idea of onion routing [20], which is widely used today. This approach depends on the use of an overlay network made up of onion routers. The onion routers are responsible for cryptographically removing each layer of a packet, one at a time, to determine the next hop routing information to eventually forward each packet to their final destinations. The weaknesses of this solution are that side channel attacks exist and have been demonstrated to be susceptible to timing attacks [21], packet counting attacks [22], and intersection attacks [23], which can all reveal the source and destination nodes of a communication stream.

The Onion Router (Tor) is one of the most popular and widely used implementations of onion routing with over 2.25 million users [24]. Tor is able to hide servers, hide services, operate over TCP, anonymize web-browsing sessions and is compatible with Socket Secure (SOCKS)-based applications for secure communications between onion routers. However, it has been shown empirically with the aid of NetFlow data that Tor traffic can be de-anonymized with accuracy rates of 81.4% [25]. The results are achieved by correlating traffic between entry and exit points within the Tor network to determine the endpoints in communication with one another. Furthermore, Tor has an overhead associated with the requirement to encrypt traffic at each of the onion routers; this overhead would need to be limited within an ICS environment to meet the real-time constraints required of these systems. Similarly, garlic routing [26] combines and anonymizes multiple messages in a single packet but is also susceptible to the same attacks.

Overlay networks have similar features as Tor but with the goal of reducing the overhead associated with a Tor network. It has been shown that overlay networks can be used to mitigate Distributed Denial of Service (DDoS) attacks [27]. The overlay networks reroute traffic through a series of hops that change over time to prevent traffic analysis. In order for users to connect to the secure overlay network, they must first know and communicate with the secure overlay access points within the network. The required knowledge of the overlay systems prevents external adversaries from attacking end hosts on the network directly. This design can be improved by relaxing the requirement of hiding the secure overlay access points within the network from the adversary. If an adversary is able to obtain the locations of the overlay access points, then the security of this implementation breaks down and is no longer effective.

Steganography is typically used to hide and covertly communicate information between multiple parties within a network. The methods described in current literature [28] include the use of IP version 4 (IPv4) header fields and reordering IPsec packets to transmit information covertly. Although the focus of the steganography research is not on anonymizing endpoints, it can be used to pass control information to aid in anonymizing network traffic. The described approach would have to be refined to increase the amount of information (e.g., $log_2(n!)$ bits can be communicated through n packets) that can be covertly communicated if significant information is desired to be exchanged. Steganography techniques have the potential to facilitate covert communication channels for MTD techniques to operate correctly but have not been applied in this fashion.

Transparently anonymizing IP-based network traffic is a promising solution that leverages Virtual Private Networks (VPNs) and the Tor service [29]. The Tor service hides a user's true IP address by making use of a Virtual Anonymous Network (VAN), while the VPN provides the anonymous IP addresses. The challenge of this solution is the requirement that every host must possess client-side software and have a VPN cryptographic key installed. In practice, it would be infeasible for this approach to scale widely, especially within ICS environments where systems cannot afford any downtime to install and maintain the VPN client-side software and the cryptographic keys that would be necessary at each of the end devices. To reduce the burden on larger-scale networks, it may be more effective to integrate this approach into the network level, as opposed to at every end device, using an SDN-based approach. In order for an MTD strategy to be effective within an ICS environment, the MTD solution must have the ability to scale to a large number of devices without significant interruptions in communications.

MTD Techniques

MTD is an active area of research that seeks to thwart attacks by invalidating knowledge that an adversary must possess to mount an effective attack against a vulnerable target [30]. For each MTD defense deployed, there is an associated delay imposed on both the adversary and on the defender of the operational system. Quantitatively analyzing the delays introduced by each additional MTD technique applied individually and in combination with one another within an ICS environment is necessary before deploying an MTD strategy. Our analysis will aid in optimally assigning the appropriate MTD techniques to enhance the overall security of a system by minimizing the operational impacts while maximizing the adversarial workload to a system. We also evaluate each of the possible MTD defenses placed at various levels of an ICS system.

MTD Categories

Because power systems are statically configured and often do not change over long periods of time, those environments are ideal for introducing and evaluating MTD-based protections. The goal of the various MTD techniques presented here is to increase the adversarial workload and level of uncertainty during the reconnaissance phases of an attack. Since it remains an open problem to completely stop a determined, well-funded patient and sophisticated adversary, increasing the delay and likelihood of detection can be an effective means of computer security. There are a variety of MTD approaches that can be categorizing according to where the defense is meant to be applied, including at the application level, the physical level, or the data level of a system. Five high-level MTD categories have been described as part of an MTD survey [31], which include dynamic platforms, dynamic runtime environments, dynamic networks, dynamic data, and dynamic software. These categories are described in the sections that follow within the context of a critical infrastructure environment.

Dynamic Platforms

PLCs, RTUs, and IEDs vary widely from one site to another within an ICS environment. There are a number of vendors that produce these end devices with different processor, memory, and communications capabilities. These devices are responsible for measuring readings from the field (such as power usage within a power grid context) and taking physical actions on a system (such as opening or closing breakers in a power grid). Many of these end devices are several decades old and must all be configured to work together. If an adversary has the ability to exploit and control these types of end devices, they would have the ability to control physical actions remotely through an attack over computer networks. At the physical layer, several MTD strategies exist to increase the difficulty of an adversary's workload to successfully exploit a system. One strategy rotates the physical devices that are activated within a system [32–34]. For this strategy to work, the physical devices and software may vary widely, but the only requirement is that they must be capable of taking in the same input and successfully producing the same output as the other devices. If there are variations in the output, then alerts can be generated to take an appropriate action. These approaches increase the difficulty from an adversary's perspective because the adversary would be required to simultaneously exploit many devices based on the same input instead of exploiting just a single device. The difficulty for the defender comes in the form of having additional devices that must be administered and managed while also ensuring the security of the monitoring agents is maintained and that they do not become additional targets themselves. Strategies such as n-variant [35] MTD techniques run several implementations of a particular algorithm with the same input where variations of outputs would be detected by a monitoring agent. Others [36] have shown firmware diversity in smart meters can limit the effectiveness of single attacks that are able to exploit a large number of devices with a single exploit. Customized exploits would have to be designed specifically for each individual device.

Dynamic Runtime Environments

Instruction set randomization (ISR) [37] and Address Space Layout Randomization (ASLR) [38] are MTD techniques that modify the execution environment of an application over time. The effectiveness of such techniques has been measured in traditional enterprise networks but has not yet been measured on devices found within ICS-based environments where real-time responses are a major requirement. The impact upon the real-time response requirement has been measured along with the adversaries increased workload when ISR [39] and ASLR [38, 40] are enabled.

Dynamic Networks

The background section describes many of the network randomization research efforts that have been performed. The resilience of network MTD techniques against several adversaries with different capabilities has been measured in prior work [41]. For network-based MTD techniques to be effective, the exact point at which the benefit of each MTD strategy to the defender is maximized and the adversarial workload is maximized must be found. The analysis should also take into account that ICS systems have strict real-time and high-availability constraints. The MTD parameters used, such as rates of randomization and the location of the MTD techniques themselves (at the network level or the end device level), to find the balance between security and usability should ensure that the solution does not hinder the operational network.

Dynamic Data

Randomizing the data within a program is another technique used to protect data stored within memory from being tampered with or exfiltrated [42]. Compiler techniques to xor memory contents with unique keys per data structure [43], randomizing Application Programming Interfaces (APIs) for an application, and Structured Query Language (SQL) string randomization [44] help protect against code injection-type attacks. These techniques have been demonstrated on web servers and have shown varying levels of impacts to the operational systems. The benefits are that adversaries can be detected if the data being randomized is accessed improperly when the system is being probed or an attack is being launched in the case of SQL string randomization.

The same techniques can be applied and measured within a control system environment to assess the feasibility of applying such techniques and meeting the real-time constraints. For example, a historian server typically maintains a database of logs within an ICS environment. This server is a prime location to apply SQL string randomization toward database accesses. Data randomization can be performed on the data stored within the registers of a PLC. To measure the effectiveness of this technique, metrics of the response times to find the delays introduced can be captured. After gathering these measurements, an evaluation of the delays introduced can be performed to ensure that they are within the acceptable limits of an ICS environment. These are a few examples of where data randomization can be applied within an ICS setting.

Dynamic Software

Introducing diversity into software implementations helps eliminate targeted attacks on specific versions of software that may be widely distributed and deployed. In the case of a widely deployed software package, compromising a single instance would then compromise the larger population of deployed instances. To introduce diversity and help prevent code injection attacks in networks, the network can be mapped to a graph coloring problem [45] where no two adjacent nodes share the same color or software implementations. This type of deployment helps prevent worms from spreading and rapidly infecting other systems in a network using a single payload. These techniques should also be considered as a defensive mechanism within an ICS environment. However, metrics and measurements need to be gathered and evaluated using software that is deployed and found within operational ICS environments.

At the instruction level, metamorphic code is another strategy that has primarily been utilized by adversaries to evade antivirus detection [46]. The code is structured so that it can modify itself continuously in time and maintain the same semantic behavior while mutating the underlying instructions of the code. The idea is similar to a quine [47] where a program is capable of reproducing itself as output. Metamorphic code reproduced semantically equivalent functionality but with an entirely new and different implementation with each replication. There are many techniques to develop metamorphic code-generating engines, but they are typically not used as a defensive strategy.

When software remains static, it becomes a dependable target that can be analyzed, tested, and targeted over long periods of time by an adversary. Introducing diversity at the instruction level helps eliminate code injection attacks and buffer overflows and limits the effectiveness of malware to a specific version of software in time. Once the code self-modifies itself, the malware may no longer be effective, depending on the self-modification being performed. Several techniques [48] exist to use self-modifying code as a defensive mechanism, such as inserting dead code, switching stack directions, substituting in equivalent but different instructions, in-lining code fragments, randomizing register allocation schemes, performing function factoring, introducing conditional jump chaining, enabling anti-debugging, and implementing function parameter reordering.

Dead Code

Dead code refers to function calls to code fragments that do not contribute to the overall goal of an algorithm and is a useful strategy to deter an adversary. Dead code fragments have the goal of causing frustration and confusion and generally wasting the time of an adversary in analyzing complex code fragments that are not of importance to the overall algorithm. However, techniques do exist to dynamically detect dead code [49] fragments, so this strategy should be deployed with care. Also of note is that if the size of a program cannot exceed a certain threshold, it may be necessary to take into account the available space on the system so that the code does not overly cause an excess amount of bloat and exceed space limitations.

Dead code can help protect against an adversary who is statically analyzing and reverse engineering a software implementation. In this case of a MTD protection, when the dead code is included, the goal is to cause the adversary to spend a significant amount of time analyzing code that is not useful to the overall software suite. This technique serves as a deterrence and a decoy to protect the important software. Dead code is often used as an obfuscation technique of software to make it more difficult for an adversary to understand [50].

Although this technique may be effective against certain types of adversaries performing static code analysis, the security is based on the assumed limited analytical and intelligence capabilities placed on an adversary. This assumption is not valid when considering nation state adversaries who have a wide array of resources in terms of finances, staff, and intelligence available. The technique also breaks down and fails when dynamic code analysis is performed to recognize that the dead code does not actually provide any contributions to the overall functionality of the software under consideration.

Stack Directions

The direction that the stack grows can be chosen to grow either at increasing memory addresses or decreasing memory addresses [51]. Buffer overflow attacks must take into account this direction to effectively overflow the return address so that the adversary can execute their own arbitrary code. One strategy to eliminate such an attack is to either run a program that dynamically selects the direction of the stack at runtime or to run two instances of a program in parallel with each instance having their stacks growing in opposite directions. The two programs would then be overseen by a monitoring agent to ensure that there are no deviations in results between the two programs. It is possible that an attack can still succeed in both cases at the same time but only in very specialized cases where the original code is written in such a way that the overflow works on different variables simultaneously in both directions that the stack grows; this is, however, unlikely to occur on the majority of code that is of practical interest to an adversary.

This technique must consider possible space limitations of the system and the overhead to detect deviations between the two versions of software. If both versions are running on the same system, then the processor utilization may also be a concern for other applications running on the system. If the implementations are on separate systems, then the network overhead to communicate the results of each run must also not negatively impact the system in question. An additional area of importance is the security of the monitoring agent to validate that a possible attack is in progress. Many security protections often become a target [52] for adversaries and need to be taken into account, as well.

Equivalent Instruction Substitution

Many techniques exist to introduce diversity into a program by substituting equivalent instructions [53]. The goal of substituting equivalent instructions for multiple instances of a program is to maintain equivalent functionality while diversifying the implementation of the underlying software. The benefit is that the difficulty of identifying functionally equivalent software implementations from one another is increased from the adversaries' perspective. This increases the difficulty placed on an adversary who is attempting to develop a scalable malware solution designed to compromise a large number of systems using a single exploit. The trade-off is typically in the increased performance of the variants of equivalent instructions. In many cases, compiler optimizers will automatically reduce high-level programming modifications to the same optimized assembly instructions using dependency graphs [54]. To disable compiler optimizations, compiler flags must be enabled to maintain the intended diversity within the binary executable. The impact on the defender and operational network of applying this approach has not been measured within an ICS environment at the time of this writing. The feasibility of applying this approach within an ICS setting will depend on the availability and number of equivalent instructions that can be replaced that function the same and do not introduce significant time delays.

MTD Applications and Scenarios Within ICS

MTD strategies can benefit a broad range of environments, spanning enterprise IT systems that are widely connected and ICS networks, which are completely isolated from the Internet. Each environment has different requirements and constraints for which the MTD approaches and parameters must be specifically configured in order for the strategy to be feasible in a practical setting. Some of the MTD parameters that can be adjusted include the frequency of reconfigurations, the amount of entropy supplied to the MTD technique when performing IP randomization, the maximum number of hops between endpoints tolerable when performing communication path randomization, and the size of a binary when performing application randomization, for example. The requirements and constraints of these systems include meeting strict performance measurements (e.g., latency, bandwidth, throughput constraints), satisfying the North American Electric Reliability Corporation (NERC) Critical Infrastructure Protection (CIP) Standards [55] and the International Organization for Standardization (ISO)/International Electrotechnical Commission (IEC) 27000 series of Information Security Management Systems (ISMS) standards [56], and conformance to the National Institute of Standards and Technology (NIST) Cybersecurity Framework [57].

Each environment has their own unique set of requirements and constraints that must be met in an operational setting. Because MTD approaches can be applied broadly across a number of environments, the parameters of the MTD strategies must be adjusted to meet the requirements and constraints of the target environment. The focus here is on ICS environments, but the approaches can be applied similarly to the other environments including enterprise networks, Internet of Things (IoT), cloud, mobile, etc.

Industrial Control Systems

The primary requirement for many of these systems is to maintain high availability and integrity [1]. In the electrical power grid, the high-availability requirement comes from the criticality of the types of systems that depend on the power grid to operate (e.g., hospitals, governments, educational institutions, commercial and residential buildings, etc.). Figure 1 shows an example of power grid and the components found at various layers of the network. These systems involve a number of utilities communicating with one another and the distribution of power across a geographically disperse area of customers. A study was performed with the goal of quantifying the economic costs associated with service interruptions to the US power grid that are estimated to be approximately $79 million annually [58]. From the 2003 blackout in New York [59], the estimated direct costs were between $4 billion and $20 billion [60] while there were also in excess of 90 deaths [61]. Though these interruptions were not due to remote attacks over computer networks, such attacks are capable of causing similar disruptions. The need for computer security within an ICS setting is clear as the impacts and consequences of downtime can be dire.

Use Case

Because ICSs operate with both legacy and modern devices, there is a mixture of serial and IP communications. Typical protocols deployed within ICS networks

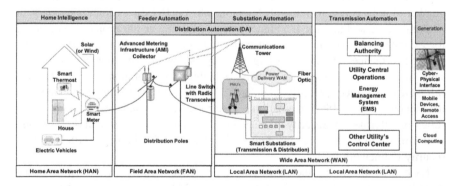

Fig. 1 An example power grid that shows the high-level components from generation of power to transmission to distribution and finally to delivery at a residential home

include Modbus [62], Distributed Network Protocol (DNP3) [63], and Process Field Net (PROFINET) [64]. These protocols are widely used within ICS environments, and many, such as Modbus, were not designed with security in mind since these protocols were originally intended for serial communications and only later expanded, with Modbus TCP, to function over IP networks. Still, the expectation was that such IP networks would be controlled and isolated. Modbus is a protocol that can be used to read and write memory values to ICS end devices, such as PLCs or industrial computers that can either sense readings from equipment or perform physical actions based on digital inputs received. Some of the physical actions include opening or closing a valve within a water pipeline, opening or closing breakers within a power system, or shutting down a power plant.

Given that ICSs are becoming more interconnected to business networks for ease of maintenance and management, remote attacks over computer networks become a real possibility since the business networks are connected to the Internet. However, as demonstrated by Stuxnet [6], a network connection to the Internet is not a requirement to exploit a system, and the attack against Home Depot [65] shows how vulnerable operational technology can be exploited to penetrate additional systems. In a scenario where the Modbus protocol is configured to read and write memory values from and to, respectively, a PLC that controls a physical process, an adversary could launch a man-in-the-middle (MITM) attack [66] to spoof values read/written to the PLC's memory. Since legacy PLCs are fundamentally different from the systems we are accustomed to working with (in terms of the memory and processing resources available) and because they were designed with the understanding that they would be used only within closed system environments, integrity and authentication checks were typically not built into these systems. As ICS environments have evolved, PLCs and other end devices are becoming more connected externally. As a result, end devices that do not have integrity and authentication checks built in are susceptible to adversaries eavesdropping on communications and/or maliciously modifying those communications via MITM attacks. To mitigate such an attack, a defender could deploy a number of strategies to protect against this threat.

If the adversary has direct access to the network and has the ability to observe or modify traffic, spoofed packets can be injected or replayed into the network. The goal of the adversary in this scenario would be to maliciously write incorrect values into a PLC's memory space to cause an unintended physical action to take place within the system. One defense that could protect against an attack, where the adversary crafts and injects packets into the network, could be to deploy an MTD strategy randomizing application port numbers in the communication channel (the Modbus standard port number is 502). Continuously changing this value in time would require the adversary to constantly track and learn the new random mappings that are active. Another defense that can be deployed would be to configure a secure communication channel between the endpoints to prevent the adversary from maliciously observing and spoofing traffic. This solution would require the adversary to compromise the underlying encryption algorithm or a cryptographic key.

The optimal solution that a defender should select depends on the capabilities of the end devices as well as the amount of delay that can be tolerated by the network. If the end devices are capable of supporting more well-established modern encryption algorithms, such as the Advanced Encryption Standard with at least a 128-bit key length (AES-128) [67], then that is the ideal solution. However, the end devices may either not be capable of supporting AES or they may not be able to afford the computationally expensive tasks, in terms of central processing unit (CPU) utilization [68], to support an encrypted channel. The amount of CPU available depends on the current load of the system. The other option is to deploy a gateway system capable of serving as a proxy to harness the necessary security protections [69–72]. This MTD approach follows the gateway solution and is capable of minimally delaying the network communications while adding on an additional layer of defense into the network. The parameters of the MTD techniques can then be adjusted to meet the criteria required by the ICS to maintain a high-availability system while avoiding the computationally expensive price of encrypting all communication channels.

Constraints

One of the major challenges for new technologies to be deployed within ICS environments is that legacy and embedded devices occupy a large portion of these systems. Some of the devices found are decades old and do not have the processing or memory resources available to harness modern security technologies. This can be attributed to the fact that many of these systems were developed starting from the 1880s to the 1930s [73] and many legacy devices are still in place today. Another constraint is that even if the devices are modern and capable of harnessing new security technologies, the software and specialized hardware are often both closed and proprietary [1]. The proprietary nature creates a challenge for security researchers to understand, integrate, and test new security protections directly into the end devices themselves. In this scenario, an additional gateway system is typically introduced to proxy the end devices with the new security technologies enabled. This proxy creates an additional hop that packets must traverse, which affects latency.

Another challenge is the diverse set of equipment that can be found within ICS environments. These devices, from multiple vendors, must interoperate with one another, which is a challenge of its own. Adding computer security protections into each of these devices directly in a vendor neutral way requires agreement and collaboration between a number of competing parties. This is a challenge that can oftentimes be the most difficult piece of the puzzle. These constraints cannot be ignored as new security technologies must be retrofitted into the existing environment with competing vendors working together, as completely replacing all of the equipment is not a valid option.

Requirements

ICSs have several requirements, regulations, and standards that must be met. Perhaps the most important requirements for ICS environments are to minimize the amount of delay introduced into a system and to ensure the integrity of the commands communicated within these environments. Latency is one of the primary metrics used and is typically constrained to 50 milliseconds and in some cases can be in the nanosecond scale [74]. Any delays on the operational network can result in instability of the power system [75]; therefore, new security protections must meet the strict time requirements to be relevant and feasible within these systems. Integrity is also a key requirement as data integrity attacks could manipulate sensors, control signals, or mislead an operator into making a wrong decision [9]. Also, interoperability requirements, as mentioned in the preceding subsection, must be met. The International Electrotechnical Commission 61850 (IEC-61850) [76] standard has outlined a general guide to achieve interoperability. To maximize the benefit of new security features introduced into an ICS system, these requirements and standards need to be met.

Experimentation

We evaluated network-based moving target defense strategies applied to a representative control system environment. The environment consisted of nine buildings with inverters communicating with a central server. The system evaluated will harness a 2.5-megawatt microgrid system, and our tests were performed before end devices were introduced into the network to evaluate the MTD strategies independently. Figure 2 shows the configuration of the network where IP randomization and port randomization were integrated. These MTD strategies continuously modified IP addresses and port numbers at varying frequencies from 1 s to completely static configurations. Our testing involved three buildings where we were able to successfully demonstrate the MTD technologies. Prior to introducing the MTD technologies, each of the three buildings included a software-defined networking (SDN)-capable switch to manage the MTD techniques. The three SDN-capable switches are shown in Fig. 2 and labeled as "SDN Switch Flow Forwarding" in the lighter blue rectangles in each of the buildings. An end device was included in Building #1 labeled as "Host 1" to represent an end device within the network. An additional Linksys router was added to the network to support an out-of-band communication channel for the SDN OpenFlow control traffic. The dashed green lines show the out-of-band communications.

To manage the randomized flows installed on the SDN switches, the "SDN Randomizer Controller" was also added to the network. This system is responsible for periodically communicating the randomized IP addresses to each of the SDN switches to create the moving target defense solution. The same "SDN Switch Flow

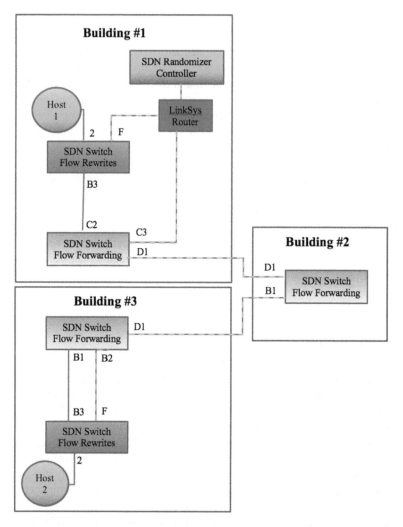

Fig. 2 Multibuilding SDN configuration with IP randomization enabled

Rewrites" SDN-capable switch was also introduced into Building #3. Additionally, a second host, labeled "Host 2," was added to Building #3 to show communications between two endpoints within the network. Building #2 only has an SDN-capable switch, labeled "SDN Switch Flow Forwarding," which has flows configured to forward both control and data traffic between Building #1 and Building #3.

"Host 1" and "Host 2" are then able to communicate with one another with IP randomization enabled. Additionally, "Host 1" and "Host 2" have machine-learning dynamic defense algorithms enabled to detect anomalous behavior on the network. Flows with the source IP address of the "SDN Randomizer Controller" are identified as management communications and are forwarded to the correct ports of the "SDN Switch Flow Rewrites," which are both shown as port "F" in the diagram.

Adversarial Scenario

We showed that if an adversary were to be introduced into the network either as (1) an insider or as (2) a result of a successfully launched MITM attack anywhere between the two "SDN Switch Flow Rewrites," then we could detect a network scan and automatically respond by randomizing the IP addresses to invalidate the information gained from the network scan. In our specific scenario, the adversary was introduced in Building #1 in the link with "B3" and "C2" as the endpoint ports. The adversary then launched the following nmap scan:

nmap -sP x.y.z.0/24

where x.y.z is the network address configured in the system. The network in this case was configured as a 24-bit network with the last 8 bits reserved for host IP addresses. The nmap command above will scan the entire IP space of the x.y.z network and report back the hosts that are active. In this scenario, once the adversary receives the results, they would then attempt to open a secure shell (ssh) session with the hosts in the network. The goal of the MTD technology is to detect the initial network scan and randomize IP addresses so that information gained about active hosts within the network would no longer be valid. To accomplish this goal, we launched the MTD strategy upon detection of the network scan. The described use case was successfully deployed within our test bed.

Metrics

Metrics were captured when the randomization schemes were performed independently and when combined with one another. The metrics collected include round-trip time, bandwidth, throughput, TCP retransmits, and dropped packets. The randomization schemes evaluated include a baseline measurement without any randomization schemes enabled, IP randomization with a frequency interval of 3 s, port randomization with a frequency interval of 3 s, and port randomization with a frequency interval of 1 s with encryption. The results are shown in Figs. 3, 4, 5, 6, 7, and 8.

Figures 3 and 4 show the results of the Internet Control Message Protocol (ICMP) round-trip times measured using the OpenDaylight controller as applied toward the scenario, as shown in Fig. 2, where "Host 1" is issuing the following command to "Host 2":

ping -c 300 host2.

In Fig. 3, each of the randomization schemes are measured independently and when combined with one another. The impacts vary slightly and are within the noise of network traffic as each scheme fluctuates outperforming and underperforming the other schemes depending on when the measurement is taken. This can be seen more clearly in Fig. 4 where the raw averages and standard deviations closely resemble one another across each of the randomization schemes. The impacts of each of the

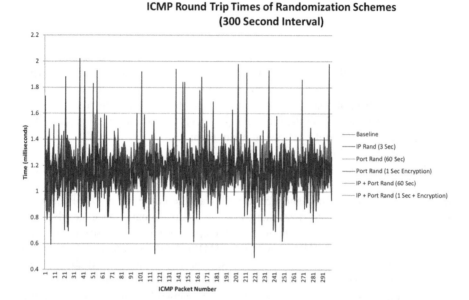

Fig. 3 ICMP round-trip time measurements when enabling each randomization scheme independently and when combined with one another over a 300-s interval

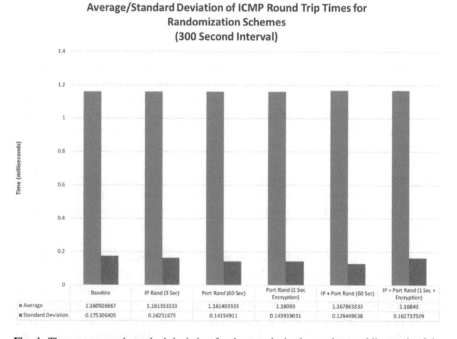

Fig. 4 The averages and standard deviation for the round-trip times when enabling each of the randomization schemes independently and in combination of each other over a 300-s interval of time

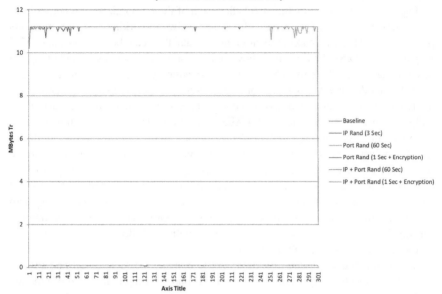

Fig. 5 Multibuilding SDN configuration with IP randomization enabled

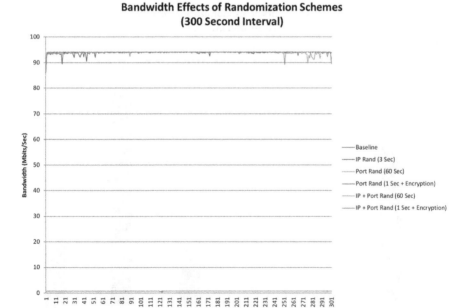

Fig. 6 The measured bandwidth when enabling each randomization scheme independently and in combination of each other over 1 s increments in a 300-s interval of time

randomization schemes in our test environment proved to be minimal from our experimental results obtained.

Figures 5 and 6 measure the effects that each of the randomization schemes have on the transfer rates and the bandwidth. Since we were working with 100 megabits per second links, the maximum amount of data that can be transferred is 12.5 megabytes (100/8). Our results show that most of the schemes, including the baseline, achieve ~11.2 megabytes of data transferred. The exceptions to this are the schemes that use the port randomization with encryption involved where ports are randomized and encrypted every second. Since we are randomizing at each of the endpoints in software and the cost for AES encryption is significant, the transfer rates were significantly impacted and yielded ~0.1 megabytes of data transferred. These results indicate that in environments where large amounts of data need to be transferred quickly, encrypting and randomizing ports every second may not be a suitable option. However, the impacts on round-trip time were minimal, so if small amounts of data are transferred, as is typically the case within control system communications, then any of the schemes may be appropriate. The metrics captured for transfer rates and bandwidth were captured with the following command:

iperf3 -c host2 -i 1 -t 300 -p 999 -V.

This command was issued for the client to connect to "Host 2" on port 99 and report back results every second over a 300-second total interval in verbose mode.

Figures 7 and 8 show the results of the number of retransmits incurred when each of the randomization schemes were enabled independently and in combination with

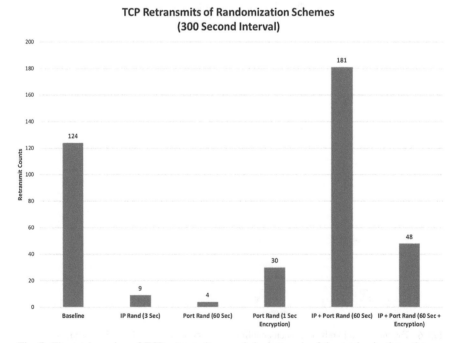

Fig. 7 The total number of TCP retransmits recorded when each of the randomization schemes were enabled independently and in combination of each other over a 300-second interval of time

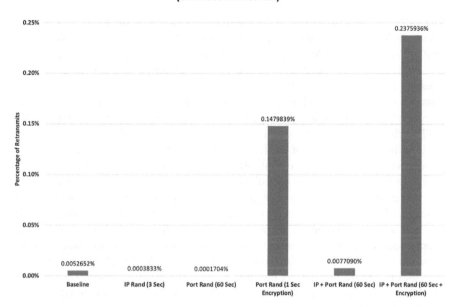

Fig. 8 The percentage of TCP retransmits recorded when each of the randomization schemes were enabled independently and in combination of each other over a 300-second interval of time

one another. When measuring the baseline configuration without any of the randomization schemes enabled, 124 retransmits were recorded or ~0.005% of the total number of packets. When enabling the IP randomization scheme and the port randomization every 60 s scheme independently, there were fewer retransmits and a fewer percentage of overall retransmits. This may be a result of a more congested network during the time that the baseline measurements were recorded. It would be expected that the baseline would be similar or better than when enabling each of the randomization schemes. The remaining three schemes all produced a higher percentage of retransmits, although only two had a higher number of total retransmits. This is due to fewer packets being transmitted when encrypting port numbers every second. Although the percentages in all cases are low, it should be considered whether these percentages are acceptable to be applied within an operational setting. The results obtained were captured using the same iperf command that was specified for the bandwidth and throughput measurements, as shown in Figs. 7 and 8. The number of dropped packets were also measured as part of these tests. All of the schemes reported no dropped packets.

Conclusion

The evaluated network-based MTD approaches have been shown to be effective within an ICS environment. We performed several experiments with a variety of configurations for each MTD technique. The techniques presented, although effective individually, are meant to be a piece of the larger computer security puzzle. These MTD techniques can be thought of as additional layers of defense to help protect a system from an adversary attempting to gain an understanding of a system in the early stages of an attack. Additional defenses can be deployed alongside the MTD techniques to create an even more secure system. Deploying an individual MTD technique or a suite of MTD techniques alongside other computer security protections will depend on the application. For example, the MTD techniques may provide a way to mitigate a "hit list" type of attack [77], but the MTD techniques themselves do not provide the ability to detect the hit list attack. Intrusion Detection Systems (IDSs), firewalls, security information and event management (SIEM) systems, and virus scanners, for example, should all be included as part of the overall security protection as well. In this scenario, we used machine-learning algorithms to detect the hit list attack and trigger the MTD schemes. The MTD strategies by themselves are not meant to be a comprehensive security solution that protects against all threats but rather should be applied as an additional layer of defense in general.

References

1. K. Stouffer, J. Falco, K. Scarfone, *Guide to Industrial Control Systems (ICS) Security*, (NIST Special Publication 800-82, 2011)
2. R. Chandia, J. Gonzalez, T. Kilpatrick, M. Papa, S. Shenoi, Security strategies for SCADA networks, in *Critical Infrastructure Protection*, (Springer, New York, 2007), pp. 117–131
3. A. Cardenas, S. Amin, Z. Lin, Y. Huang, C. Huang, S. Sastry, Attacks against process control systems: Risk assessment, detection, and response, in *Proceedings of the 6th ACM Symposium on Information, Computer and Communications Security*, (ACM, 2011), pp. 355–366
4. Y. Huang, A. Cardenas, S. Amin, Z. Lin, H. Tsai, S. Sastry, Understanding the physical and economic consequences of attacks on control systems. Int. J. Crit. Infr. Prot. **2**(3), 73–83 (2009)
5. B. Miller, D. Rowe, A survey of SCADA and critical infrastructure incidents, in *Proceedings of the 1st Annual Conference on Research in Information Technology*, (ACM, 2012), pp. 51–56
6. N. Falliere, L. Murchu, E. Chien, W32. Stuxnet Dossier, in *White paper, Symantec Corp., Security Response*, vol. 5, (2011)
7. P. Pourbeik, P. Kundur, C. Taylor, The anatomy of a power grid blackout. IEEE Power. Energ. Mag. **4**(5), 22–29 (2006)
8. G. Liang, S. Weller, J. Zhao, F. Luo, Z. Dong, The 2015 Ukraine blackout: Implications for false data injection attacks. IEEE Trans. Power. Syst. **32**, 3317–3318 (2017)
9. S. Sridhar, M. Govindarasu, Data integrity attacks and their impacts on SCADA control system, in *IEEE PES General Meeting*, (July 2010), pp. 1–6
10. H. Farhangi, The path of the smart grid. IEEE Power. Energ. Mag. **8**(1), 18–28 (2010)
11. R. Robles, M. Choi, E. Cho, S. Kim, G. Park, J. Lee, Common threats and vulnerabilities of critical infrastructures. Int. J. Control. Autom. **1**(1), 17–22 (2008)

12. C. Hauser, D. Bakken, A. Bose, A failure to communicate: Next generation communication requirements, technologies, and architecture for the electric power grid. IEEE Power. Energ. Mag. **3**(2), 47–55 (2005)
13. R. Rajkumar, I. Lee, L. Sha, J. Stankovic, Cyber-physical systems: The next computing revolution, in *Design Automation Conference*, (June 2010), pp. 731–736
14. G. Ericsson, Cybersecurity and power system communication essential parts of a smart grid infrastructure. IEEE Trans. Power. Deliver. **25**(3), 1501–1507 (2010)
15. S. Hofmeyr, S. Forrest, Architecture for an artificial immune system. Evol. Comput. **8**(4), 443–473 (2000)
16. E. Al-Shaer, Q. Duan, J. Jafarian, Random host mutation for moving target defense, in *SecureComm*, (Springer, 2012), pp. 310–327
17. S. Antonatos, P. Akritidis, E. Markatos, K. Anagnostakis, Defending against hitlist worms using network address space randomization. Comput. Netw. **51**(12), 3471–3490 (2007)
18. K. Farris, G. Cybenko, Quantification of moving target cyber defenses, in *SPIE Defense+ Security*, (International Society for Optics and Photonics, 2015), pp. 94560L–94560L
19. A. Sharon, R. Levy, Y. Cohen, A. Haiut, A. Stroh, D. Raz, Automatic network traffic analysis, 24 Oct 2000, U.S. Patent 6,137,782
20. D. Goldschlag, M. Reed, P. Syverson, Onion routing for anonymous and private internet connections. Commun. ACM **42**(2), 39–41 (1999)
21. V. Shmatikov, M. Wang, Timing analysis in low latency mix networks: Attacks and defenses, in *ESORICS 2006. LNCS*, ed. by D. Gollmann, J. Meier, A. Sabelfeld, vol. 4189, (Springer, Heidelberg, 2006), pp. 18–33
22. J. Raymond, Traffic analysis: Protocols, attacks, design issues, and open problems, in *Designing Privacy Enhancing Technologies, Lecture Notes in Computer Science, LNCS 2009*, ed. by H. Federath, (Springer-Verlag, 2001), pp. 10–29
23. R. Dingledine, N. Mathewson, P. Syverson, Tor: The second-generation onion router, in *Usenix Security*, (2004)
24. *The Tor Project*, (2014), URL https://metrics.torproject.org/torperf.html
25. S. Chakravarty, M.V. Barbera, G. Portokalidis, M. Polychronakis, A. Keromytis, On the effectiveness of traffic analysis against anonymity networks using flow records, in *PAM*, (Springer-Verlag New York, New York, 2014), pp. 247–257
26. G. Tchabe, Y. Xu, *Anonymous Communications: A Survey on I2P*, (CDC Publication Theoretische Informatik-Kryptographie und Computeralgebra, 2014), URL https://www.cdc. informatik.tu-darmstadt.de
27. A. Keromytis, V. Misra, D. Rubenstein, SOS: An architecture for mitigating DDoS attacks. IEEE J. Sel. Area. Comm. **22**(1), 176–188 (2004)
28. K. Ahsan, D. Kundur, Practical data hiding in TCP/IP, in *Proceedings of the Workshop on Multimedia Security at ACM Multimedia*, vol. 2, (2002)
29. L. Pimenidis, T. Kolsch, Transparent anonymization of IP-based network traffic, in *Proceedings of the 10th Nordic Workshop on Secure IT-Systems*, (2005)
30. S. Jajodia, A. Ghosh, V. Subrahmanian, V. Swarup, C. Wang, X. Wang (eds.), *Moving Target Defense II – Application of Game Theory and Adversarial Modeling* (Springer, New York, 2013)
31. H. Okhravi, M. Rabe, T. Mayberry, W. Leonard, T. Hobson, D. Bigelow, W. Streilein, *Survey of Cyber-Moving Target Techniques*, (Massachusetts Institute of Technology, Lexington Lincoln Laboratory Technical Report, 2013)
32. B. Salamat, T. Jackson, G. Wagner, C. Wimmer, M. Franz, Runtime defense against code injection attacks using replicated execution. IEEE Trans. Depend. Secure. Comput. **8**(4), 588–601 (2011)
33. D. Holland, A. Lim, M. Seltzer, An architecture a day keeps the hacker away. ACM SIGARCH Comp. Arch. News. **33**(1), 34–41 (2005)
34. H. Okhravi, A. Comella, E. Robinson, J. Haines, Creating a cyber-moving target for critical infrastructure applications using platform diversity. Int. J. Crit. Infr. Prot. **5**(1), 30–39 (2012)

35. B. Cox, D. Evans, A. Filipi, J. Rowanhill, W. Hu, J. Davidson, J. Knight, A. Nguyen-Tuong, J. Hiser, N-variant systems: A secretless framework for security through diversity, in *USENIX Security Symposium*, (2006), pp. 105–120

36. S. McLaughlin, D. Podkuiko, A. Delozier, S. Miadzvezhanka, P. McDaniel, Embedded firmware diversity for smart electric meters, in *HotSec'10 Proceedings of the 5th USENIX Conference on Hot Topics in Security*, (2010)

37. G. Kc, A. Keromytis, V. Prevelakis, Countering code-injection attacks with instruction-set randomization, in *Proceedings of the 10th ACM conference on Computer and Communications Security*, (2003), pp. 272–280

38. H. Shacham, M. Page, B. Eu-Jin Goh, N. Modadugu, D. Boneh, On the effectiveness of address-space randomization, in *Proceedings of the 11th ACM Conference on Computer and Communications Security*, (2004), pp. 298–307

39. A. Sovarel, D. Evans, N. Paul, Where's the FEEB? The effectiveness of instruction set randomization, in *USENIX Security Symposium*, (2005)

40. J. Ganz, S. Peisert, ASLR: How robust is the randomness? in *Proceedings of the 2017 IEEE Secure Development Conference (SecDev)*, (2017)

41. A. Chavez, W. Stout, S. Peisert, Techniques for the dynamic randomization of network attributes, in *Proceedings of the 49th Annual International Carnahan Conference on Security Technology*, (2015)

42. S. Forrest, A. Somayaji, D. Ackley, Building diverse computer systems, in *The Sixth Workshop on Hot Topics in Operating Systems*, (IEEE, 1997), pp. 67–72

43. C. Cadar, P. Akritidis, M. Costa, J. Martin, M. Castro, Data randomization, Microsoft Research Technical Report TR-2008-120, (2008)

44. S. Boyd, A. Keromytis, SQLrand: Preventing SQL injection attacks, in *Applied Cryptography and Network Security*, (Springer, Berlin/Heidelberg, 2004), pp. 292–302

45. A. O'Donnell, H. Sethu, On achieving software diversity for improved network security using distributed coloring algorithms, in *Proceedings of the 11th ACM Conference on Computer and Communications Security*, (2004), pp. 121–131

46. Q. Zhang, D. Reeves, Metaaware: Identifying metamorphic malware, in *Twenty-Third Annual Computer Security Applications Conference, ACSAC 2007*, (IEEE, 2007), pp. 411–420

47. M. Rieback, B. Crispo, A. Tanenbaum, Is your cat infected with a computer virus? in *Fourth Annual IEEE International Conference on Pervasive Computing and Communications*, (2006), pp. 10

48. I. You, K. Yim, Malware obfuscation techniques: A brief survey, in *2010 IEEE International Conference on Broadband, Wireless Computing, Communication and Applications*, (2010), pp. 297–300

49. J. Butts, G. Sohi, Dynamic dead-instruction detection and elimination. ACM SIGOPS Operat. Syst. Rev. **36**(5), 199–210 (2002)

50. A. Sung, J. Xu, P. Chavez, S. Mukkamala, Static analyzer of vicious executables (SAVE), in *20th Annual IEEE Computer Security Applications Conference*, (2004), pp. 326–334

51. B. Salamat, A. Gal, M. Franz, Reverse stack execution in a multi-variant execution environment, in *Workshop on Compiler and Architectural Techniques for Application Reliability and Security*, (2008), pp. 1–7

52. B. Min, V. Varadharajan, U. Tupakula, M. Hitchens, Antivirus security: Naked during updates. Softw. Pract. Exper. **44**(10), 1201–1222 (2014)

53. B. De Sutter, B. Anckaert, J. Geiregat, D. Chanet, and K. Bosschere. Instruction set limitation in support of software diversity, in *Information Security and Cryptology ICISC 2008*, ed. by P. Lee, J. Cheon, vol. 5461 of Lecture Notes in Computer Science, (Springer, Berlin/Heidelberg, 2009), pp. 152–165

54. D. Kuck, R. Kuhn, D. Padua, B. Leasure, M. Wolfe, Dependence graphs and compiler optimizations, in *Proceedings of the 8th ACM SIGPLAN-SIGACT Symposium on Principles of Programming Languages*, (1981), pp. 207–218

55. North American Electricity Council (NERC), *Critical Infrastructure Protection (CIP) Reliability Standards*, (2009), URL http://www.nerc.com/pa/Stand/Pages/CIPStandards.aspx
56. G. Disterer, ISO/IEC 27000, 27001 and 27002 for information security management. J. Infor. Secur. **4**(2), 92 (2013)
57. National Institute of Standards and Technology (NIST), Cybersecurity Framework (CSF), (2014), URL https://www.nist.gov/cyberframework
58. K. LaCommare, J. Eto, Cost of power interruptions to electricity consumers in the United States (U.S.). Energy **31**(12), 1845–1855 (2006)
59. G. Andersson, P. Donalek, R. Farmer, N. Hatziargyriou, I. Kamwa, P. Kundur, N. Martins, J. Paserba, P. Pourbeik, J. Sanchez-Gasca, R. Shultz, J. Stankovic, C. Taylor, V. Vittal, Causes of the 2003 major grid blackouts in North America and Europe, and recommended means to improve system dynamic performance. IEEE Trans. Power. Syst. **20**(4), 1922–1928 (2005)
60. U.S.-Canada Power System Outage Task Force, Final Report on the August 14, 2003, blackout in the United States and Canada: Causes and recommendations, Merrimack Station AR-1165. URL https://www3.epa.gov/region1/npdes/merrimackstation/pdfs/ar/AR-1165.pdf
61. G. Anderson, M. Bell, Lights out: Impact of the August 2003 power outage on mortality in New York, NY. Epidemiology **23**(2), 189 (2012)
62. IDA Modbus, *Modbus Application Protocol Specification v1. 1a*, (North Grafton, Massachusetts, 2004), URL www.modbus.org/specs.php
63. G. Clarke, D. Reynders, E. Wright, *Practical Modern SCADA Protocols: DNP3, 60870.5, and Related Systems*, (Newnes, 2004)
64. F. Joachim, PROFINET-scalable factory communication for all applications, in *Proceedings of Factory Communication Systems*, (IEEE, 2004), pp. 33–38
65. R. Walters, Cyber-attacks on U.S. companies in 2014, in *The Heritage Foundation*, vol. 4289, (2014), pp. 1–5
66. B. Zhu, A. Joseph, S. Sastry, A taxonomy of Cyber-attacks on SCADA systems, in *Internet of Things (iThings/CPSCom), 4th IEEE International Conference on Cyber, Physical and Social Computing*, (2011), pp. 380–388
67. F. Miller, A. Vandome, J. McBrewster, *Advanced Encryption Standard*, (2009)
68. P. Chodowiec, Comparison of the hardware performance of the AES candidates using reconfigurable hardware, Ph.D. thesis, (George Mason University, 2002)
69. F. Robertson, J. Carroll, W. Sanders, T. Yardley, E. Heine, M. Hadley, D. McKinnon, B. Motteler, J. Giri, W. Walker, E. McCartha, *Secure Information Exchange Gateway for Electric Grid Operations*, (Grid Protection Alliance Technical Report, Chattanooga, TN, 2014)
70. S. Hurd, J. Stamp, A. Chavez, *OPSAID Initial Design and Testing Report*, (Department of Energy, 2007)
71. B. Smith, J. Stewart, R. Halbgewachs, A. Chavez, Cybersecurity interoperability: The Lemnos project, in *53rd ISA POWID Symposium*, vol. 483, (2010), pp. 50–59
72. R. Halbgewachs, A. Chavez, OPSAID improvements and capabilities report, Sandia National Laboratories Technical Report, (2011)
73. T. Hughes, *Networks of Power: Electrification in Western Society, 1880–1930* (JHU Press, Baltimore, 1993)
74. P. Castello, P. Ferrari, A. Flammini, A. Muscas, S. Rinaldi, An IEC 61850-compliant distributed PMU for electrical substations, in *Applied Measurements for Power Systems (AMPS), 2012 IEEE International Workshop*, (2012), pp. 1–6
75. F. Milano, M. Anghel, Impact of time delays on power system stability. IEEE Trans. Circuits. Syst. I Reg. Papers. **59**(4), 889–900 (2012)
76. R. Mackiewicz, Overview of IEC 61850 and benefits, in *IEEE Power Systems Conference and Exposition, PSCE'06*, (2006), pp. 623–630
77. S. Staniford, V. Paxson, N. Weaver, How to own the Internet in your spare time, in *USENIX Security Symposium*, vol. 2, (2002), pp. 14–15

Proactive Defense Through Deception

Massimiliano Albanese and Sushil Jajodia

Abstract Cyberattacks are typically preceded by a reconnaissance phase in which attackers aim at collecting valuable information about the target system, including network topology, service dependencies, operating systems (OSs), and unpatched vulnerabilities. Unfortunately, when system configurations are static, given enough time, attackers can always acquire accurate knowledge about the target system through a variety of tools—including OS and service fingerprinting—and engineer effective exploits. To address this important problem and increase the resiliency of systems against known and unknown attacks, many techniques have been devised to dynamically and periodically change some aspects of a system's configuration in order to introduce uncertainty for the attacker. However, these techniques, commonly referred to as moving target defenses, may introduce a significant overhead for the defender. To address this limitation, we present a graph-based approach for manipulating the attacker's view of a system's attack surface, which does not require altering the actual configuration of a system. To achieve this objective, we first formalize the notions of system view and distance between views and then define a principled approach to manipulating responses to attacker's probes so as to induce an external view of the system that satisfies certain desirable properties. In particular, we propose efficient algorithmic solutions to two classes of problems, namely, (i) inducing an external view that is at a minimum distance from the internal view while minimizing the cost for the defender and (ii) inducing an external view that maximizes the distance from the internal view, given an upper bound on the cost for the defender. In order to demonstrate practical applicability of the proposed approach, we present deception-based techniques for defeating an attacker's effort to fingerprint OSs and services on the target system. These techniques consist in manipulating outgoing traffic so that it resembles traffic generated by a completely different system. Experimental results show that our approach can efficiently and effectively deceive an attacker.

The work presented in this chapter was partially supported by the Army Research Office under grant W911NF-13-1-0421.

M. Albanese (✉) · S. Jajodia
Center for Secure Information Systems, George Mason University, Fairfax, VA, USA
e-mail: malbanes@gmu.edu; jajodia@gmu.edu

© Springer Nature Switzerland AG 2019

C. Rieger et al. (eds.), *Industrial Control Systems Security and Resiliency*, Advances in Information Security 75, https://doi.org/10.1007/978-3-030-18214-4_9

169

Introduction

Today's approach to cyber-defense is governed by slow and deliberative processes, such as deployment of security patches, testing, episodic penetration exercises, and human-in-the-loop monitoring of security events. Adversaries can greatly benefit from this situation and can continuously and systematically probe target networks with the confidence that those networks will change slowly, if at all. In fact, cyberattacks are typically preceded by a reconnaissance phase in which adversaries aim at collecting valuable information about the target system, including network topology, service dependencies, operating systems (OSs) and applications, and unpatched vulnerabilities. As most system configurations are static—hosts, networks, software, and services do not reconfigure, adapt, or regenerate except in deterministic ways to support maintenance and uptime requirements—it is only a matter of time for attackers to acquire accurate knowledge about the target system. While these issues are common across industries and applications, their potential consequences may be more critical within the domain of industrial control systems where additional factors—such as concerns about compatibility among the many cyber and physical components—may contribute to further slowing down the process of upgrading software and deploying patches. A vast array of automated tools and techniques exist to facilitate reconnaissance efforts, including OS and service fingerprinting tools. Specifically, OS fingerprinting techniques are designed to determine the OS of a remote host either in a passive way, through sniffing and traffic analysis, or in an active way, through probing. Similarly, service fingerprinting aims at determining what services are running on a remote host. The information collected during reconnaissance will eventually enable attackers to engineer reliable exploits and plan attacks. A taxonomy of OS fingerprinting tools is presented in section "Fingerprinting".

In order to address this important problem, significant work has been done in the area of adaptive cyber-defense (ACD), which includes concepts such as moving target defense (MTD), artificial diversity, and bioinspired defenses. Essentially, a number of techniques have been proposed to dynamically change a system's attack surface by periodically reconfiguring some aspects of the system. In Manadhata and Wing [1], a system's attack surface has been defined as the "subset of the system's resources (methods, channels, and data) that can be potentially used by an attacker to launch an attack." Intuitively, dynamically reconfiguring a system is expected to introduce uncertainty for the attacker and increase the cost of the reconnaissance effort. However, one of the major drawbacks of current approaches is that periodically reconfiguring a system may introduce significant overhead for the defender and for legitimate users, as well as the potential for Denial of Service conditions, because reconfiguring a system might make it temporarily unavailable. Additionally, most of the existing techniques are purely proactive in nature or do not adequately consider the attacker's behavior.

The work we present in this chapter advances the state of the art in adaptive cyber-defense by developing a graph-based approach for manipulating the attacker's perception of a system's attack surface. To achieve this objective, we formalize the notion of *system view* as well as the notion of *distance between views*. We refer to the attacker's view of the system as the *external view* and to the defender's view as the *internal view*. A system's attack surface can then be thought of as the subset of the internal view that would be exposed to potential attackers when no deceptive strategy is adopted. Starting from these definitions, we develop a principled, yet practical, approach to manipulate outgoing traffic so as to induce an external view of the system that satisfies certain desirable properties. In particular, we propose efficient algorithmic solutions to two different classes of problems, namely, (i) inducing an external view that is at a minimum distance from the internal view while minimizing the cost for the defender and (ii) inducing an external view that maximizes the distance from the internal one, given an upper bound on the cost for the defender.

Our approach goes beyond simply introducing uncertainty for the attacker and deceives potential intruders into making incorrect inferences about important system characteristics, including OSs and active services. In order to demonstrate practical applicability of the proposed approach, we present deception-based techniques for defeating an attacker's effort to fingerprint OSs and services on the target system. Differently from many existing techniques, we do so without changing the actual configuration of the system. In fact, our approach mainly consists in manipulating outgoing traffic such that not only important details about OSs and services are not revealed, but network traffic also resembles traffic generated by hosts and networks with different characteristics. Experiments conducted on a prototypical implementation show that the overhead introduced by the proposed approach is negligible, thus rendering this solution completely transparent to legitimate users. At the same time, our approach can effectively deceive the attackers and steer them away from critical resources we wish to protect.

The remainder of the chapter is organized as follows. Section "Related Work" discusses related work. Section "Threat Model" discusses the threat model, whereas section "Motivating Example" presents a motivating example. Next, section "Deception Approach" provides a detailed description of our approach and presents the problem statement, as well as the proposed algorithms. Then, section "Fingerprinting" presents the details of specific techniques we have designed to defeat OS and service fingerprinting. Finally, section "Experimental Evaluation" reports the results of our experiments, and section "Conclusion" gives some concluding remarks.

Related Work

Moving target defense (MTD) defines mechanisms and strategies to increase complexity and cost for attackers [2]. MTD approaches aiming at selectively altering a system's attack surface usually involve reconfiguring the system in order to make an attack's preconditions unstable, thus preventing the attack from succeeding.

Dunlop et al. [3] propose a mechanism to dynamically hide addresses of IPv6 packets to achieve anonymity. This is done by adding virtual network interface controllers and sharing a secret among all the hosts in the network. Duan et al. [4] present a proactive Random Route Mutation technique to randomly change the route of network flows to defend against eavesdropping and DoS attacks. In their implementation, they use OpenFlow Switches and a centralized controller to define the route of each flow. Jafarian et al. [5] use an IP virtualization scheme based on virtual DNS entries and software-defined networks. Their goal is to hide network assets from scanners. Using OpenFlow, each host is associated with a range of virtual IP addresses and mutates its IP address within its pool. A similar identity virtualization approach is presented in Albanese et al. [6]. In chapter "Moving Target Defense to Improve Industrial Control System Resiliency" of Jajodia et al. [7], an approach based on diverse virtual servers is presented. Each server is configured with a set of software stacks, and a rotational scheme is employed for substituting different software stacks for any given request, thus creating a dynamic and uncertain attack surface. Also, Casola et al. [8, 9] propose an MTD approach for protecting resource-constrained distributed devices through fine-grained reconfiguration at different layers of the software stack.

These solutions reconfigure a system in order to modify its external attack surface. On the other hand, the external view of the system is usually inferred by attackers based on the results of probing and scanning tools. Starting from this observation, our approach consists in modifying system responses to probes in order to expose an external view of the system that is significantly different from the actual attack surface, without altering the system itself.

Reconnaissance tools, such as nmap or Xprobe2, can identify a service or an OS by analyzing packets that can reveal implementation-specific details about the target host [10, 11]. Network protocol fingerprinting refers to the process of identifying specific features of a network protocol implementation by analyzing its input/output behavior [12]. These features may reveal specific information such as protocol version, vendor information, and configurations. Reconnaissance tools store known system's features and compare them against scan responses in order to match a fingerprint. Watson et al. [11] adopted protocol scrubbers in order to avoid revealing implementation-specific information and restrict an attacker's ability to determine the OS of a protected host. Moreover, some proof-of-concept software and kernel patches have been proposed to alter a system fingerprint [13], such as IP Personality and Stealth Patch. Among the various techniques that have been proposed to defeat fingerprinting [14], a very simple and intuitive one consists in modifying the default values of a TCP/IP stack implementation, such as the TTL, window size, or TCP Options. Other approaches to defeating fingerprinting [15] include altering public service banners and searching content files for *incriminating* strings that can give away information about the OS. For instance, web pages may include automatically generated comments that identify the authoring tool.

Honeypots have been traditionally used to try to divert attackers away from critical resources. Although our approach and honeypots share a common goal, they are significantly different. Our approach does not alter the system while honeypot-based solutions introduce vulnerable machines in order to either capture the attacker [16] or collect information for forensic purposes [17]. Instead, we aim at deceiving attackers by manipulating their view of the target system and forcing them to plan attacks based on inaccurate knowledge so that the attacks will likely fail. To the best of our knowledge, we have been the first to propose an adaptive and comprehensive approach to changing the attacker's view of a system's attack surface without reconfiguring the system itself [18].

Threat Model

We assume that an external adversary is attempting to infer a detailed view of the target network, using reconnaissance tools, such as nmap [10], to discover active hosts in the network along with their configurations. The information an attacker may attempt to discover includes OSs, exposed services and their version, network topology, and routing information. The attacker will then leverage this knowledge to plan and execute attacks aimed at exploiting exposed services. We also assume that the attacker may use OS fingerprint techniques consisting in sending valid and invalid IP packets and analyzing the respective responses. Moreover, we limit service fingerprinting to the case of TCP probes, as it is the case for most common probing tools.

The attacker's strategy is illustrated by the flowchart in Fig. 1. The goal is to launch an attack against one of the hosts in the target network. Multiple stages of this attack strategy (marked with a red cross in the figure) can be defeated using our approach. For instance, we may expose services with no exploitable vulnerabilities, or for a given service, we may expose exploitable vulnerabilities which do not correspond to actual vulnerabilities of that service.

Motivating Example

As a reference scenario, we consider the networked system of Fig. 2, modeling the IT infrastructure of a fictitious e-commerce company. Customers can access publicly available services through a website hosted in the DMZ. The business logic and critical services are deployed in the intranet, and some of these services need to be accessible through the Internet in order to allow company branches to process orders and query the inventory.

Our goal is to modify the attacker's view of the system. In order to do so, we only modify system-dependent information exposed by system-specific protocol implementations. We adopt a graph-based strategy to generate different views of the system, such as the one in Fig. 3, by repeatedly applying *view manipulation primitives*, which are implemented by filtering and altering outgoing traffic.

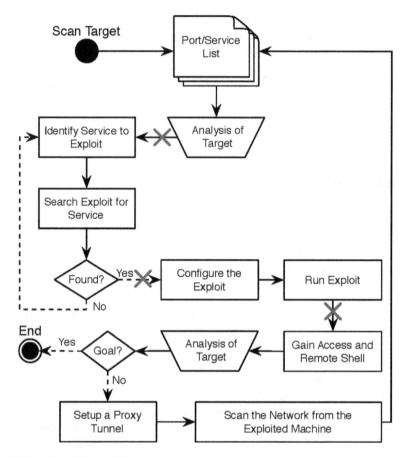

Fig. 1 Flowchart of the attacker strategy

In Fig. 3, we depict machines whose configuration has been manipulated with a different color or texture compared to Fig. 2. A change in the OS is represented by a different letter on the top right corner of each machine. In this example, applying several manipulation steps to the original view, we move from a scenario where all the servers have the same OS to a scenario in which each OS is different from the real one and the others. Similarly, all of the services are altered. For instance, we alter the database server fingerprint so as it will be recognized as an implementation from a different vendor. As for the public web server, we want it to act like two web servers in a load-balancing configuration. To do so, we mutate, with a certain frequency, both the OS and the service fingerprints and modify the packet-level parameters. In this way, we can force the attacker to believe that multiple servers need to be compromised in order to disrupt the service.

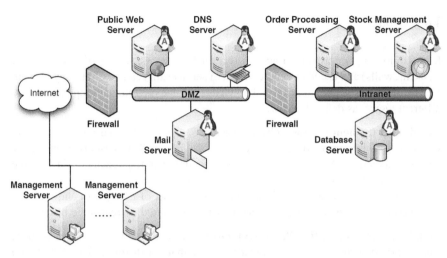

Fig. 2 Flowchart of the attacker strategy

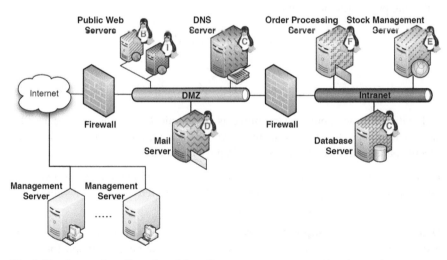

Fig. 3 Topology and configuration of the reference system as presented to the attacker

Deception Approach

In order to achieve our goal of inducing an attacker's view of the system's attack surface that is measurably different from the internal view, we first formalize the notions of *view*, *manipulation primitive*, and *distance between views*.

View Model

In the following, we assume a system is a set $S = \{s_1, s_2, \ldots, s_n\}$ of devices (e.g., hosts, firewalls) and use Ψ to denote the set of services that can be offered by hosts in S. The defender's and attacker's knowledge of the system are represented by different views, as defined below.

Definition 1 (System's View) Given a system S, a *view* of S is a triple $V = S_o, C, \nu$, where $S_o \subseteq S$ is a set of observable devices, $C \subseteq S_o \times S_o$ represents connectivity between elements in S_o, and $\nu : S_o \to 2^{\Psi}$ is a function that maps each host in S_o to the set of services it offers.

Intuitively, a view represents knowledge of a subset of the system and includes information about the topology and about services offered by reachable hosts.[1]

Definition 2 (Manipulation Primitive) Given a system S and a set $\mathcal{V}$ of views of S, a *manipulation primitive* is a function $\pi : \mathcal{V} \to \mathcal{V}$ that transforms a view $V' \in \mathcal{V}$ into a view $V'' \in \mathcal{V}$. Let Π denote a family of such functions. For each $\pi \in \Pi$, the following properties must hold:

$$(\forall V \in \mathcal{V})\pi(V) \neq V$$

$$(\forall V \in \mathcal{V})(\nexists \langle \pi_1, \pi_2, \ldots, \pi_m \rangle \in \Pi^m \mid \pi_1(\pi_2(\ldots(\pi_m(V))\ldots)) = \pi(V))$$

Intuitively, a manipulation primitive is an atomic transformation that can be applied to a view to obtain a different view. Each primitive may have a set of specific parameters, which we omit to simplify the notation.

Example 1 A possible manipulation primitive is π_{OS_B}, which transforms a view V' into a view V'' by changing the OS fingerprint of a selected host. Figure 4 illustrates the effect of this primitive on the system's view.

Definition 3 (View Manipulation Graph) Given a system S, a set $\mathcal{V}$ of views of S, and a family Π of manipulation primitives, a view manipulation graph for S is a directed graph $G = (\mathcal{V}', \mathcal{E}, \ell)$, where:

- $\mathcal{V}' \subseteq \mathcal{V}$ is a set of views of S
- $\mathcal{E} \subseteq \mathcal{V} \times \mathcal{V}$ is a set of edges
- $\ell : \mathcal{E} \to \Pi$ is a function that associates with each edge (V', V'') manipulation primitive $\pi \in \Pi$ such that $V'' = \pi(V')$

The node representing the internal view has no incoming edges. All other nodes represent possible external views.

---

[1]A more complete definition of view could incorporate information about service dependencies and vulnerabilities, similarly to what is proposed in Jafarian et al. [5]

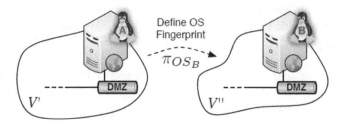

Fig. 4 Example of manipulation primitive

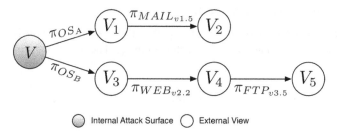

○ Internal Attack Surface ○ External View

Fig. 5 Example of a view manipulation graph

Figure 5 shows an example of a view manipulation graph. After applying any π, a new view is generated. By analyzing the graph, one can enumerate all possible ways to generate external views starting from the internal view V of a system S.

Definition 4 (Distance) Given a system S, a set $\mathcal{V}$ of views of S, a *distance* over $\mathcal{V}$ is a function $\delta : \mathcal{V} \times \mathcal{V} \to \mathbb{R}$ such that, $\forall V', V'', V''' \in \mathcal{V}$, the following properties hold:

$$\delta(V', V'') \geq 0$$

$$\delta(V', V'') = 0 \Leftrightarrow V' = V''$$

$$\delta(V', V'') = \delta(V'', V')$$

$$\delta(V', V''') \leq \delta(V', V'') + \delta(V'', V''')$$

In the simplest case, the distance can be measured by looking at the elements that differ between views. To do so, we can consider the difference in the number of hosts between two views. Then, for each host that is present in both views, we can add one if OS fingerprints differ and we can add one if service fingerprints differ. More sophisticated distances can be defined, but this is beyond the scope of the discussion in this chapter.

Definition 5 (Path Set) Given a view manipulation graph G, the *path set* P_G for G is the set of all possible paths (sequence of edges) in G. We denote the k-th path of length m_k in P_G as $p_k = \left\langle \pi_{i_1}, \pi_{i_2}, \ldots, \pi_{i_{m_k}} \right\rangle$.

Example 2 Consider the view manipulation graph G in Fig. 5. The set of all possible paths is $P_G = \{\langle \pi_{OSA} \rangle, \langle \pi_{MAILv1.5} \rangle, \langle \pi_{OSA}, \pi_{MAILv1.5} \rangle, \langle \pi_{OSB} \rangle, \langle \pi_{WEBv2.2} \rangle$, $\langle \pi_{FTPv3.5} \rangle, \langle \pi_{OSB}, \pi_{WEBv2.2} \rangle, \langle \pi_{WEBv2.2}, \pi_{FTPv3.5} \rangle, \langle \pi_{OSB}, \pi_{WEBv2.2}, \pi_{FTPv3.5} \rangle\}$.

We will use the notation $V_a \xrightarrow{p_k} V_b$ to refer to a path p_k originating from V_a and ending in V_b. For instance, in Fig. 5, the path that goes from V to V_5 is denoted as $V \xrightarrow{p_9} V_5 = \langle \pi_{OSB}, \pi_{WEBv2.2}, \pi_{FTPv3.5} \rangle$.

Definition 6 (Cost Function) Given a path set P_G, a *cost function* is a function $f_c : P_G \to \mathbb{R}$ that associates a cost to each path in P_G. The following properties must hold:

$$f_c(\langle \pi_i \rangle) \geq 0$$

$$f_c(\langle \pi_{i_j}, \pi_{ij+1} \rangle) \geq \min \left(f_c(\langle \pi_{ij} \rangle), f_c(\langle \pi_{ij+1} \rangle) \right)$$

$$f_c(\langle \pi_{i_j}, \pi_{ij+1} \rangle) \leq f_c(\langle \pi_{ij} \rangle) + f_c(\langle \pi_{ij+1} \rangle)$$

If the third property above holds strictly, then f_c is said to be additive.

Problem Statement

We can now formalize the two related problems we are addressing in this chapter:

Problem 1 Given a view manipulation graph G, its internal view V_i, and a distance threshold $d \in \mathbb{R}$, find an external view V_d and a path $V_i \xrightarrow{p_b} V_d$ that minimizes $f_c(p_d)$ subject to $\delta(V_i, V_d) \geq d$.

Problem 2 Given a view manipulation graph G, its internal view V_i, and a budget $b \in \mathbb{R}$, find an external view V_b and a path $V_i \xrightarrow{p_b} V_b$ that maximizes $\delta(V_i, V_b)$ subject to $f_c(p_b) \leq d$.

Algorithms

In this section, we present heuristic algorithms to solve the problems defined in section "Problem Statement". The algorithms start from the internal view and explore the state space by iteratively traversing the k most promising outgoing edges of each node traversed until a termination condition is reached. To quantify the benefit of traversing a given edge, we define a benefit score as the ratio of the distance between the corresponding views to the cost for achieving that distance. For each node in the graph, we only traverse the k outgoing edges with the highest values of the benefit score.

Algorithm *TopKDistance*

To solve Problem 1, we first generate the view manipulation graph and then execute the heuristic *TopKDistance* algorithm. For efficiency purposes, we only generate a subgraph G_d of the complete-view manipulation graph G, such that generation along a given path stops when the distance from the internal view becomes equal to or larger than the minimum required distance d. In fact, any additional edge would increase the cost of the solution; thus, it would not be included in the optimal path.

Algorithm 1 describes how we generate the subgraph G_d. The algorithm uses a queue Q to store the vertices to be processed. At each iteration of the `while` loop (Line 3), a vertex V is popped from the queue and the maximum distance from the internal view V_o is updated (Line 4). The constant MAX_INDEGREE (Line 5) is used to test if a node has been fully processed. When the in-degree is equal to MAX_INDEGREE, the node has been linked to all the nodes that differ by only one element. In this case, there are no new vertices that can be generated starting from this node. Given the set $C(s)$ of all admissible configurations for a device $s \in S$, MAX_INDEGREE can be computed as MAX_INDEGREE $= \sum_{s \in S} (|C(S)| - 1)$

Algorithm 1 *GenerateGraph(V_o, C, d)*

Input: The internal view V_o, the set of admissible configurations C for each host, the minimum required distance d.
Output: A subgraph G_d of the complete view manipulation graph with vertices within distance d of V_o.
1: // Initialization: Q is the queue of vertices to be processed (initially empty); G_d is also initially empty.
2: $maxDistance \leftarrow 0$; $G_d.addVertex(V_o)$; $Q.push(V_o)$
3: **while** $Q \neq 0$ **do**
4:     $V \leftarrow Q.pop()$; $maxDistance \leftarrow \max(maxDistance, \delta(V_o, V))$;
5:     **if** $V.indegree <$ MAX_INDEGREE **then**
6:         $predecessors \leftarrow G_d.getPredecessors(V)$
7:         $newVertices \leftarrow getCombinations(V, predecessors, C)$
8:         **for all** $V_n \in newVertices$ **do**
9:             $G_d.addVertex(V_n)$
10:             $G_d.addDirectedEdge(V, V_n, cost(V, V_n), \delta(V, V_n))$ // Add an edge from V to V_n
11:             $verticesToLink \leftarrow getOneChangeVertices(G_d, V_n)$
12:             **for all** $V' \in verticesToLink$ **do**
13:                 $G_d.addBidirectEdge(V, V_n, cost(V_n, V), \delta(V_n, V))$ // Add a bidirectional edge between V_n and V
14:             **end for**
15:             **if** $maxDistance \leq d$ **then**
16:                 $Q.push(V_n)$
17:             **end if**
18:         **end for**
19:     **end if**
20: **end while**

On Line 6, the set of V's predecessors is retrieved. The function *getCombinations* (Line 7) generates all possible configurations that differ from V's configuration by only one element. Both V and its predecessors are excluded from the returned array. The function *getOneChangeVertices* (Line 11) returns an array of all the vertices whose configuration differs just for one element from the vertex given as input. Those vertices need to be linked with the vertex V (Lines 12–14). Lines 15–17 check if the maximum distance d has been reached. If this is not the case, the newly generated vertex is pushed into the queue to be examined.

Once the subgraph has been generated, we can run the top-k analysis using the *TopKDistance* algorithm (Algorithm 2), which recursively traverses the subgraph to find a solution. We use V to denote the vertex under evaluation in each recursive call. Line 1 creates an empty list to store all the paths discovered from V. Line 2 is one of the two termination conditions. It checks whether the current distance is greater than or equal to d or no other nodes can be reached. The second term in the termination condition takes into account both the case of a node with no outgoing edges and the case of a node whose successors are also its predecessors. We do not consider edges directed to predecessors in order to construct loop-free paths. If the termination condition is satisfied, a solution has been found and a path from V to V can be constructed by closing the recursion stack. Line 3 creates an empty path and adds V to it. Then, the path list is updated and returned. On Line 6, all the edges originating from V are sorted by decreasing benefit score. Then, Lines 7–13 perform the top-k analysis and *TopKDistance* is recursively invoked for each of the best k destinations. The result is a list of paths having vertex V as the origin (Lines 10–12).

Algorithm 2 *TopKDistance(G, V, k, d, cc, cd)*

Input: A graph G, a vertex V, an integer k, a minimum distance d, the current cost cc, and the current distance cd.
Output: A list of paths *pathList*.

1: $pathList \leftarrow \emptyset$
2: **if** $cd \geq d \vee V.successors \setminus (V.successors \cap V.predecessors) = \emptyset$ **then**
3:      $p \leftarrow emptyPath$; $p.addVertex(V)$; $pathList \leftarrow \{p\}$
4:      **return** $pathList$
5: **end if**
6: $sort(V.outgoingEdges)$; $numToEval \leftarrow \min(k, |V.outgoingEdges|)$
7: **for all** $i \in [1, numToEval]$ **do**
8:      $V_n \leftarrow V.outgoingEdges[i].getDestination()$
9:      $eval \leftarrow TopKDistance(G, V_n, k, d, update(cc), update(cd))$
10:     **for all** $p \in eval$ **do**
11:         $p.addVertex(V)$; $p.addDirectedEdge(V, V_n, \delta(V, V_n), cost(V, V_n))$ // Add an edge from V to V_n
12:     **end for**
13: **end for**
14: **return** $pathList$

Algorithm *TopKDistance*

Algorithm *TopKBudget* (Algorithm 3) implements both graph generation and exploration in order to improve time efficiency in the resolution of Problem 2. Our approach is to generate the graph only in the k most promising directions in order to limit graph generation. The algorithm uses a queue to store examined paths that may represent a solution.

Algorithm 3 *TopKBudget(G, V_o, C, b)*

Input: A graph G, the internal view V_o, the set of admissible configurations C for each host, and a budget $b \in \mathbb{R}$.
Output: A list of paths.
1:   // Initialization: Q is the queue of paths to process (initially empty); *solutions* is initially empty; G is a graph and contains only V_o
2:   $p \leftarrow emptyPath;\ p.addVertex(V_o);\ Q.add(p)$
3:   **while** $Q \neq \emptyset$ **do**
4:       $dataToExplore \leftarrow \emptyset$
5:       $p \leftarrow Q.pop();\ v \leftarrow p.getLast()$
6:       **if** $V.indegree <$ MAX_INDEGREE **then**
7:           $predecessors \leftarrow G.getPredecessors(V)$
8:           $newVertices \leftarrow getCombinations(V, predecessors, C)$
9:           **for all** $V_n \in newVertices$ **do**
10:             $G.addVertex(V_n)$
11:             $G.addDirectedEdge(V, V_n, cost(V, V_n), \delta(V, V_n))$ // Add an edge from V to V_n
12:             $verticesToLink \leftarrow getOneChangeVertices(G, V_n)$
13:             **for all** $V' \in verticesToLink$ **do**
14:                 $G.addBidirectEdge(V, V_n, cost(V_n, V), \delta(V_n, V))$ // Add a bidirectional edge between V_n and V'
15:             **end for**
16:         **end for**
17:     **end if**
18:     **for all** $V_n \in V.getDirectSuccessors()$ **do**
19:         **if** $\neg V.isPrecedessor(V_n) \land \neg p.contains(V_n) \land (p.totalCost + cost(V, V_n) \leq b)$ **then**
20:             $importance \leftarrow \delta(V, V_n)/cost(V, V_n) + estimate(V_n, p, C)$
21:             $dataToExplore.add([V_n, p, importance])$
22:         **end if**
23:     **end for**
24:     $sort(dataToExplore)$
25:     $numToEval \leftarrow \min(k, dataToExplore.size)$
26:     **if** $numToEval = 0$ **then**
27:         $solutions \leftarrow p$
28:     **end if**
29:     **for all** $i \in [1, numToEval]$ **do**
30:         $newPath \leftarrow dataToExplore[i].p$
31:         $newPath.addAsLeaf(V_n)$
32:         $Q.push(newPath)$
33:     **end for**
34: **end while**
35: **return** *solutions*

Line 5 retrieves a path p from the queue and its last vertex V. Lines 6–17 perform graph generation starting from V. The generation process is similar to the one described for Algorithm 1. On Line 18, all the successors of V (generated or linked at this stage) have been computed. All the successor nodes are used to compute the importance level of V. We sum V's benefit (distance/cost) and an importance estimation of V's successors. This estimation provides some knowledge about the solution we may discover by further exploring from V. It is done by the function estimate, which returns the value of the maximum benefit of vertices reachable from V's successor V. A triple including the importance level, the successor V_n, and the path p under examination is then added to a list (Lines 20–21).

On Line 24, the list is sorted by decreasing importance level. Line 26 checks if there is no further exploration to perform. In this case, the path under examination is added to the list of solutions. Otherwise (Lines 29–32), new paths are generated from p. Each of the paths is p plus a new node that is in the top k successors of V. All the newly generated paths are then pushed into the queue for further examination. When the queue becomes empty, the complete list of solutions is returned.

Fingerprinting

Operating system (OS) fingerprinting is the practice of determining the OS of a remote host on a network. This may be accomplished either passively, by sniffing and analyzing network packets traveling between hosts, or actively, by sending carefully crafted packets to the target host and analyzing the responses [15]. Active fingerprinting approaches are typically more sophisticated than passive fingerprinting. In the simplest scenario, the attacker does not resort to stealth techniques and gathers information about the OS by trying to connect to the host. For instance, while establishing a connection via the standard Telnet or SSH protocol, the OS version is often sent to the client as part of a welcome message. Moreover, some FTP server implementations allow to retrieve this information through the SYST command. In general, active fingerprinting techniques trigger the target into sending one or more responses, which are then analyzed by the attacker to infer the type and version of the OS installed on the remote host. Carefully crafted ICMP, TCP, and UDP packets are sent to the target in order to observe how it responds to both valid and invalid packets. For instance, in the case of TCP probes, features that can be used to distinguish between different OSs include (i) the relative order of the TCP Options and (ii) the total length of TCP Options. Additionally, the IP header may reveal some information about the nature of the OS.

Conversely, passive fingerprinting consists in using a packet sniffer to passively collect and analyze packets traveling between hosts. A simple passive method consists in inspecting the Time To Live (TTL) field in the IP header and the TCP window size of the SYN or SYN + ACK packet in a TCP session. The values of both the initial TTL and the TCP window size depend on the specific OS implementation, as shown in Table 1. One reason for this is that RFC specifications define intervals of values and recommended values but do not mandate specific values. For instance, RFC 1700 recommends to initialize TTL to 64. Of course, relying only on the TTL value is not sufficient to determine the OS because, given the nature of this parameter, the TTL decreases as a packet traverses the network, and inferring the correct initial TTL may not always be possible.

Many different fingerprinting tools are available today. To better assess the impact of our approach, we have defined a taxonomy to classify existing fingerprinting tools

Table 1 OS-dependent IP and TCP parameters

Operating system	IP initial TTL	TCP window size
Linux Kernel 2.4/2.6	64	5840
Windows XP	128	65,535
Windows 7	128	8192

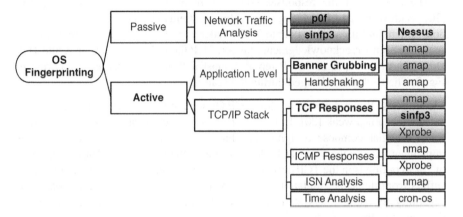

Fig. 6 Deception in the overall context of perception

based on the different approaches they implement. Figure 6 shows the proposed taxonomy. Given the variety of existing tools, it is practically impossible to develop a single technique that would defeat all of them. However, the proposed approach is effective against the most widely used fingerprinting tools. The tools that we explicitly target in our work are shown in boldface, whereas other tools that are at least partially impacted by our deception techniques are shown with a colored background. In the following sections, we provide a detailed description of some of the tools that we explicitly target in our work, namely, SinFP3, pOf, and Nessus. We refer the reader to Albanese et al. [19] for a description of additional tools.

Sin FP3

The development of SinFP was prompted by the need to reliably identify a remote host's OS under worst-case network conditions: (i) only one TCP port is open; (ii) traffic to all other TCP and UDP ports is dropped by filtering devices; (iii) stateful inspection is enabled on the open port; and (iv) filtering devices with packet normalization are in place [20]. In this scenario, only standard probe packets that use the TCP protocol can reach the target and elicit a response packet.

SinFP uses three probes: the first probe P1 is a standard packet generated by the connect () system call, the second probe P2 is the same as P1 but with different TCP Options, and the third probe P3 has no TCP Options and has the TCP SYN and

ACK flags set. The first two probes elicit two TCP SYN + ACK responses from the target. The third probe has the objective of triggering the target into sending a TCP RST + ACK response. After the three probes have been sent and the three replies have been received, a signature is built from the analysis of the response packets. Then, a signature-matching algorithm searches a database for a matching OS fingerprint.

The analysis of the responses considers both IP headers and TCP headers [20]. With respect to IP headers, the following fields are analyzed: TTL, ID, and don't fragment bit. With respect to TCP headers, the following fields are analyzed: sequence number, acknowledgment number, TCP flags, TCP window size, and TCP Options. An example of SinFP report against a Windows 7 target is reported in Fig. 7. The packets exchanged during the scan are shown in Fig. 8.

Limitations When there are too few TCP Options in P2's response, the signature's entropy becomes weak [20]. In fact, TCP Options are the most discriminant characteristics that compose a signature. That is because virtually no two systems implement exactly the same TCP Options, nor in the same order. Thus, when only the MSS option is in the TCP header, the risk of misidentification is high. SinFP also suffers from the same limitation of all knowledge-based fingerprinting tools: their capability to identify a system is limited by the existence of a corresponding fingerprint in the database.

```
...
score 100: Windows: Microsoft: Windows: Vista (RC1)
score 100: Windows: Microsoft: Windows: Server 2008
score 100: Windows: Microsoft: Windows: 7 (Ultimate)
score 100: Windows: Microsoft: Windows: Vista
...
```

Fig. 7 SinFP report against a Windows 7 host

Fig. 8 Packets exchanged during a SinFP scan

pOf

pOf (v3) is a tool that utilizes an array of sophisticated, purely passive traffic fingerprinting mechanisms to identify the players behind any incidental TCP/IP communication [21]. Its fingerprint contains:

1. **ver**: IP protocol version
2. **ittl**: Initial TTL used by the OS
3. **olen**: Length of IPv4 options or IPv6 extension headers
4. **mss**: Maximum segment size (MSS), if specified in TCP Options
5. **wsize**: Window size, expressed as a fixed value or a multiple of MSS, of MTU, or of some random integer
6. **scale**: Window scaling factor, if specified in TCP Options
7. **olayout**: Comma-delimited layout and ordering of TCP Options, if any. Supported values: explicit end of options, no-op option, maximum segment size, window scaling, selective ACK permitted, time stamp
8. **quirks**: Comma-delimited properties observed in IP or TCP headers
9. **pclass**: Payload size

Limitation The initial TTL value is often difficult to determine since the TTL value of a sniffed packet will vary depending on where it is captured. The sending host will set the TTL value to the OS's default initial TTL value, but this value will then be decremented by one for every router the packet traverses on its way to the destination. An observed IP packet with a TTL value of 57 can therefore be expected to be a packet with an initial TTL of 64 that has done 7 hops before it was captured. This tool also suffers from the TCP Options entropy issue described for SinFP.

Nessus

Nessus provides a comprehensive analysis of a target, including information about its OS and vulnerabilities. Tenable Research introduced a highly accurate form of OS identification [22]. This method combines the outputs of various other plug-ins that execute separate techniques to guess or identify a remote OS. It is worth noting that some of these techniques could also be adopted independently by an attacker. Each of these plug-ins reports a confidence level for their scan results. An example of Nessus output for OS identification is reported in Fig. 9.

Limitation Nessus's approach to fingerprinting can be very effective when used during a "credentialed" scan. Otherwise, it will report partial information, and in some cases, it will not use all the plug-ins it is equipped with. Additionally, Nessus's approach to identify vulnerabilities is strictly dependent on service banners and welcome messages. Generally, Nessus merely checks if the service's version present in the service's banner belongs to a certain interval. For instance, if a vulnerability is known to be present in a service up until version 2.0, it is really

```
The remote host is running Linux Kernel 2.4
Confidence Level : 70
Method : SinFP
The remote host is running Linux Kernel 2.6
Confidence Level : 60
Method : ICMP
```

Fig. 9 Example of Nessus OS identification report

simple to make Nessus generate false negatives by exposing a fake service banner claiming that the service version is higher than 2.0.

Fingerprint Manipulation

With respect to OS fingerprinting, our approach to deceive attackers relies on modifying outgoing traffic in a way that such traffic resembles traffic generated by a different protocol stack implementation. As we pointed out in section "Fingerprinting", protocol specifications may leave some degrees of freedom to developers. The choices that a developer makes with respect to (i) default values (e.g., initial TTL, size of the TCP window), (ii) length of TCP Options, or (iii) order of the TCP Options may reveal the nature of the OS or even the type of device (e.g., firewall, switch, router, printer, or general-purpose machine).

All the information required to impersonate a certain OS or device can be extracted from the SinFP's signature database. All the outgoing packets that may reveal relevant information about the OS are modified to reflect the deceptive signature, as shown in Fig. 10.

The most critical step in this process is the manipulation of the TCP Options. In order to present the attacker with a deceptive signature, not only do we need to modify some parameters, but we also need to reorder the options and correctly place no-operation[2] option codes to obtain the right options length. The TCP Options format can be inferred from the signature. Modifying the options length requires to adjust the total length field in the IP header and the offset value in the TCP header and subsequently adjust the sequence numbers. On the bright side, the majority of commonly used OSs share the same length for TCP Options.

With respect to service fingerprinting, we need to consider the following two cases: (i) the service banner can be modified through configuration files; and (ii) the service banner is hardcoded into the service executable. Being able to modify the packet carrying the identifying information before it leaves the host (or the network) enables to successfully address both scenarios. Moreover, even if services are under

---

[2] As specified in RFC 793, this option code may be used between options, for example, to align the beginning of a subsequent option on a word boundary.

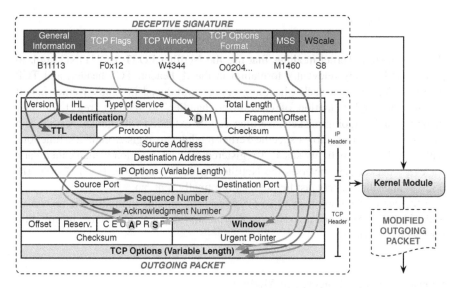

Fig. 10 Manipulation of outgoing traffic to reflect deceptive signatures

our control, we prefer to alter service banners in a completely transparent way. Our long-term goal is to develop a network appliance that can be deployed at the network boundary and is able to transparently manipulate services and OS fingerprints.

It is worth noting that when the original service banner is replaced with a deceptive banner of a different length, we need to (i) adjust the size of the packet based on the difference in length between the two banners, (ii) modify the total length field in the IP header, (iii) modify the sequence numbers in order to be consistent with the actual amount of bytes sent, and (iv) correctly handle the case of fragmented packets, which requires to reassemble a packet before modifying it.

However, this approach is not applicable to all categories of services. Services that actively use the banner information during the connection process (such as SSH) require us to use a nontransparent approach. For instance, the SSH protocol actively uses the banner information while generating hashes in the connection phase. The banner format is "SSH-protocolversion-softwareversion comments \r\n". Even though this approach can deceive tools like nmap and amap, modifying the banner will cause legitimate user to receive termination messages from the server[3] such as (i) Bad packet length or (ii) Hash Mismatch.

In summary, defeating passive tools requires to modify all outgoing packets, whereas defeating active tools only requires to alter those packets that are likely to be part of an attacker's probes.

---

[3]Errors occurs only during the connection phase, and altering the banner will not affect previously established connections.

Implementation

We implemented kernel modules that use the Netfilter POST_ROUTING hook to process and modify relevant information in the IP header, TCP header, and TCP payload. Netfilter is a packet-handling engine introduced in Linux Kernel 2.4. It enables the implementation of customized handlers to redirect, reject, or alter incoming and outgoing packets. Netfilter can be extended with hooks. A hook is a function handler that allows specific kernel modules to register callback functions within the kernel's network stack. A registered callback function is then called back for every packet that traverses the Netfilter stack. We used the POST_ROUTING hook to alter the packets just before they are finally sent out.

Specifically, we implemented an *OS fingerprint module* to modify the responses to the SinFP's probes and a *service fingerprint module* to modify banner information for specific services.

Operating System Fingerprint Module

The hook function checks if the packet is a response to the first SinFP's probe (P1): an ACK + SYN packet with a length of 44 bytes (in the case of an underlying Linux Kernel 3.02). If this is the case, the packet is altered in order to mimic a particular OS (more details are provided later); otherwise, the module checks whether the packet is an ACK + SYN with a length of 60 bytes (in the case of an underlying Linux Kernel 3.02). This packet is used in most TCP connections, and it might be a response to the second SinFP's probe (P2). If so, the packet is modified accordingly based on the target OS fingerprint.

Additionally, we verified that this approach can deceive the p0f tool when we modify the TTL value of all IP packets and the window size value of all TCP packets. During the packet manipulation stage, we track whether any of the following has been altered: the IP header, the TCP header, the length of TCP Options, and the TCP payload or its size. Based on this information, we (i) modify the IP total length value, if the size of the TCP payload has changed; (ii) recompute the TCP offset value in the TCP header and the IP total length, if the length of the TCP Options has changed; (iii) recompute the TCP checksum, if the TCP header and/or the TCP payload have been altered; and (iv) recompute the IP checksum, if the IP header has been altered.

In order to modify the responses such that they appear to have been generated by a specific OS, we created a script that (i) extracts the required characteristics of the responses to the first and second probe from SinFP's signature database and (ii) generates the C code necessary to alter the responses. The script determines how the following policies should be implemented:

- **ID policy**: the ID could be a fixed value different from zero, zero, or a random number.
- **Don't fragment bit policy**: the DF bit can be enabled or disabled.

- **Sequence number policy**: the sequence number can be zero or not altered.
- **Ack number policy**: the ack number can be zero or not altered.
- **TCP flags policy**: the TCP flags value is copied from the signature.
- **TCP window size policy**: the window size is copied from the signature.
- **TCP MSS Option policy**: the MSS value is copied from the signature.
- **TCP WScale Option policy**: the WScale is copied from the signature.
- **TCP Options policy**: the TCP Options layout is copied from the signature.

The generated code is then compiled in order to build the actual kernel module. The scheme of the resulting kernel module is presented in Listing 1 below. We assume that all the set and get functions are able to access the packet and track if the IP or TCP header has been modified.

Service Fingerprint Module

In order to alter the service fingerprint, we modify the banner sent by the application either at the time of establishing a connection or in the header of each application-level protocol data unit. Packets matching the service source port one wants to protect are analyzed. If a packet contains data, the banner string is searched and subsequently replaced. When replacing the banner, the packet size can vary: the packet is then resized according to the specific case. Listing 2 shows the sample pseudo-code for the case of an Apache Server.[4]

Experimental Evaluation

In this section, we report the results of the experiments we conducted to validate the proposed approach. We evaluated the performance of algorithms *TopKDistance* (section "Evaluation of TopKDistance") and *TopKBudget* (section "Evaluation of TopKBudget") in terms of processing time and approximation ratio for different numbers of hosts and different numbers of admissible configurations. We also evaluated our approach for deceiving fingerprinting tools from the point of view of both legitimate users interacting with the system (section "Legitmate User Perspective") and attackers trying to determine the OS of a remote host or the type of services running on it (section "Drawbacks and Limitations").

---

[4]For the sake of brevity, we omit the code for checksum recomputation.

```
if (ip->protocol == TCP && ip->len == 44 && tcp->ack == 1 && tcp->syn == 1) {
    // Probably 1st sinfp3's probe Response
    set_id();
    set_df_bit();
    set_ttl();
    set_tcp_window();
    set_tcp_flags();
    set_tcp_sequence();
    set_tcp_ack();

    if(new_option_len != option_len) {
        modify_packet_size(); // expands or shrinks packet and updates IP Length and Offset
    }

    set_tcp_options(MSS, WScale, Option_Layout);
}
else if (ip->protocol == TCP && ip->len == 60 && tcp->ack == 1 && tcp->syn == 1) {
    // Probably 2nd sinfp3's probe Response
    // Extract the timestamp from the packet and save it for re-injecting it in the right
    // position later
    timestamp = get_tcp_timestamp();

    set_id();
    set_df_bit();
    set_ittl();
    set_tcp_window();
    set_tcp_flags();
    set_tcp_sequence();
    set_tcp_ack();

    if (new_option_len != option_len) {
        modify_packet_size(); //expands or shrinks packet and updates IP Length and Offset
    }
    set_tcp_options(timestamp, MSS, WScale, Option_Layout);
}

if (tcpHeader_modified) {
    tcp->check = 0;
    tcp->check = tcp_csum();
}

if (ipHeader_modified) {
    ip->check = 0;
    ip->check = ip_csum();
}
```

Listing 1 OS deception kernel module

```
#define FAKE_APACHE_BANNER "Apache/1.1.23"
...
if (ntohs(tcph->source) == 80 && len > 0) {
    // Pointers to where to store start/end addresses of the Apache Banner String for
    // substitution
    char *b = NULL, *l = NULL;

    // Pointer to the TCP payload
    char *p= (char *)((char *)tcph+(uint)(tcph->doff*4));

    b = strstr(p, "\r\nServer:");   //String Search
    if (b != NULL) l = strstr(((char *)b + 10), "\r\n");

    if (b != NULL && l != NULL) {
        // b points to \r\nServer: x, so we add 10 to move to the beginning of x
        uint8_t signature_len = l - (b + 10);

        if (signature_len != (sizeof(FAKE_APACHE_BANNER)-1)) {
            resize_packet();
        }
        copy(b + 10, FAKE_APACHE_BANNER, sizeof(FAKE_APACHE_BANNER)-1);
    }
    ...
}
```

Listing 2 Service deception kernel module

Evaluation of TopKDistance

First, we show that, as expected, the processing time increases when the number of admissible configurations for each host increases. Figure 11 shows processing time as a function of the number of hosts for $k = 7$ and a required minimum distance $d = 5$. The processing time is practically linear in the number of hosts in the case of 3 configurations per host, but as soon as the number of configurations increases, it becomes polynomial, as shown in the case of 5 configurations per host.

Figure 12 shows the processing time as a function of the graph size for different values of k. The graph size is measured as the number of nodes that have a distance from the internal view that is less than or equals to d.

Comparing the trends for $k \in [3, 5]$, one can see that the algorithm is polynomial for $k = 3$ and linear for $k \in [4, 5]$. This can be explained considering the fact that for $k = 3$, it is necessary to explore the graph in more depth than in the case of $k \in [4, 5]$. Moreover, if we consider values of k bigger than 5, the trend is again polynomial due to the fact that the algorithm starts exploring the graph more broadly. Indeed, as we will show shortly, relatively small values of k provide a good trade-off between approximation ratio and processing time; therefore, this result is extremely valuable. To better visualize the relationship between processing time and k, we plotted the average processing time against k (see Fig. 13). The trend can be approximated by a polynomial function and the minimum is between $k = 2$ and $k = 3$. For k greater than 4, the average time to process the graph increases almost linearly.

Moreover, we evaluated the approximation ratio achieved by the algorithm. To compute the approximation ratio, we divided the cost of the algorithm's solution by the optimal cost. In order to compute the optimal solution, we exhaustively measured the shortest path (in term of cost) from the internal view to all the solutions with a

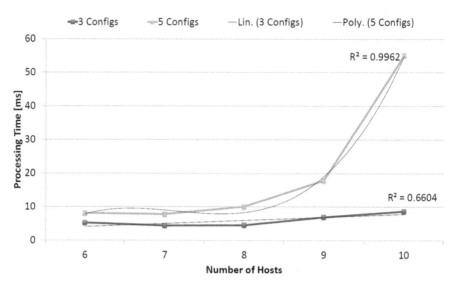

Fig. 11 Processing time vs. number of hosts

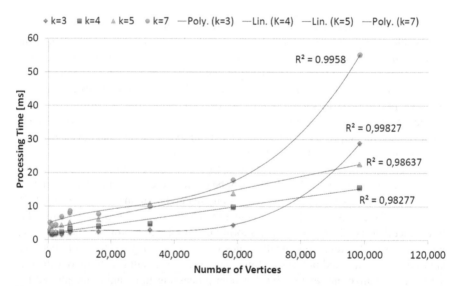

Fig. 12 Processing time vs. graph size

distance greater than the minimum required d and sorted those results by increasing
cost. The optimal solution has the maximum distance and the minimum cost. When
the algorithm could not find a solution (none of the discovered paths has a distance
greater than the minimum required d), we considered an infinite approximation.
Figure 14 shows how the ratio changes when k increases in the case of a fixed

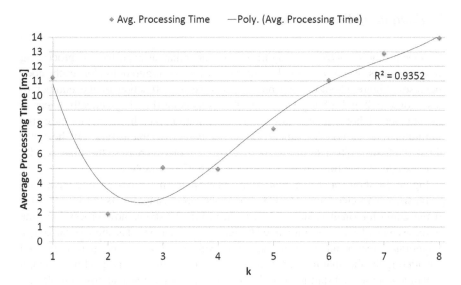

Fig. 13 Average processing time vs. *k*

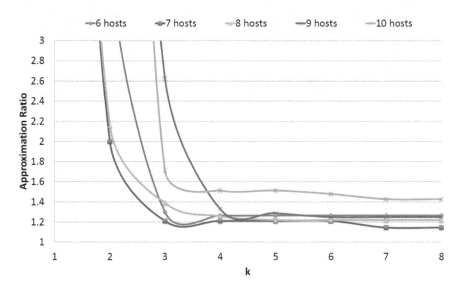

Fig. 14 Approximation ratio vs. *k*

number of configurations per node (5 in this case) and for increasing numbers of hosts. It is clear that the approximation ratio improves when k increases. Relatively low values of k (between 3 and 6) are sufficient to achieve a reasonably good approximation ratio in a time-efficient manner.

Evaluation of TopKBudget

As done for the *TopKDistance* algorithm, we show that, as expected, the processing time increases when the number of admissible configurations for each node increases. Figure 15 shows the processing time as a function of the number of hosts for $k = 2$ and a budget $b = 18$. The processing time is practically linear in the number of hosts in the case of 3 configurations per host. In this case, the minimum time (6 hosts) is about 150 ms and the maximum time (10 hosts) is about 3500 ms. When the number of configurations increases, the time rapidly increases due to the time spent in the generation of the graph.

Figure 16 shows a scatter plot of average processing times against increasing graph sizes. This chart suggests that, in practice, processing time is linear in the size of the graph for small values of k. Similarly, Fig. 17 shows how processing time increases when k increases for a fixed budget $b = 18$. The trend is approximated by a polynomial function and tends to saturate for values of $k \geq 6$. This can be explained considering the fact that for larger values of k, most of the time is spent in the graph generation phase, and starting from $k = 6$, the graph is generated almost completely. Even in this case, the important result is that small values of k achieve linear time. Moreover, for these values, the algorithm can achieve a good approximation ratio.

To compute the approximation ratio, we divided the optimal distance by the distance returned by the algorithm. In order to compute the optimal solution, we exhaustively measured the shortest path (in term of distances) from the internal view to all the solutions in a given graph. Due to the fact that it would be unfeasible to generate an exhaustive graph, we generated a subgraph up to a maximum number of nodes. We then ordered the paths by decreasing values of the distance and noted the

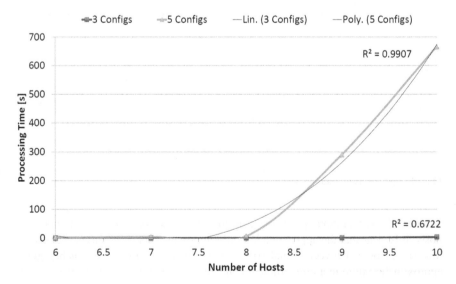

Fig. 15 Processing time vs. number of hosts

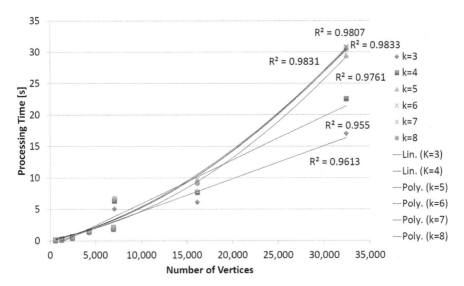

Fig. 16 Processing time vs. graph size

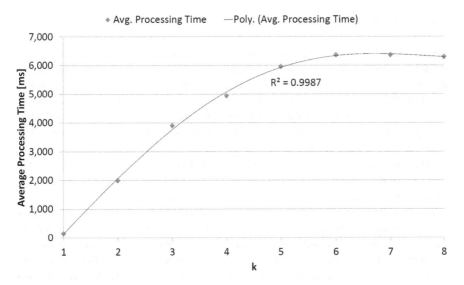

Fig. 17 Average processing time vs. k

cost needed to reach the solution. We then started the algorithm with a budget equal to this cost. Figure 18 shows how the ratio changes when k increases for a fixed number of configurations per node (5 in this case) and for increasing numbers of hosts. The approximation ratio is good even for $k = 1$, but a more accurate solution can be obtained for $k \in [2, 3]$. Larger values of k are not ideal in terms of time efficiency.

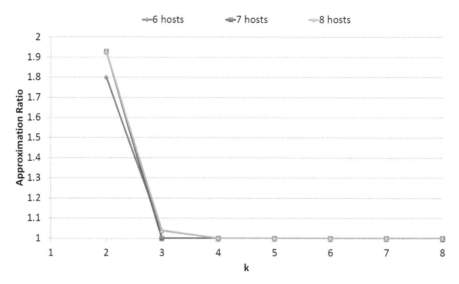

Fig. 18 Approximation ratio vs. k

Legitmate User Perspective

In the next set of experiments, we evaluated our approach for deceiving fingerprinting tools from the point of view of legitimate users interacting with the system being
defended. Our goal is to make manipulation of outgoing traffic completely transparent to users from both a functional and a performance perspective. To this end, we
ran performance tests with the Apache Benchmark, testing the server's ability to
process 20,000 requests, with a maximum of 200 simultaneous active users. The
results are shown in Fig. 19.

The tests we performed involved different system configuration scenarios: (i) the
behavior of the system is not altered (original); (ii) the kernel module to alter the OS
fingerprint and deceive only active fingerprinting tools is enabled (Sinfp3); (iii) the
kernel module to alter the service fingerprint is enabled (Apache); (iv) both modules
from scenarios (ii) and (iii) are enabled (Sinfp3 + Apache); (v) the kernel module to
alter the OS fingerprint and deceive both active and passive fingerprinting tools is
enabled (Sinfp3 + p0f); and (vi) both modules from scenarios (iii) and (v) are
enabled. The performance degradation for scenario (ii) is negligible as only two
packets need to be altered for each connection. On the other hand, when the OS
fingerprint kernel module alters all the outgoing packets (scenario (v) above), there is
a slight delay in the response time due to the larger number of packets that needs to
be altered. When the Service Fingerprint Kernel Module is enabled (scenario (iii)
above), the response time increases due to the string comparison operations
performed to identify and replace the banner information. It is clear from Fig. 19
that the Service Fingerprint Kernel Module has the largest impact on the performance of the system. However, even in the worst-case scenario, the performance
degradation is limited.

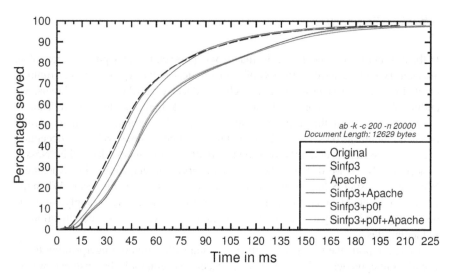

Fig. 19 Apache Benchmark

Fig. 20 Apache
Benchmark

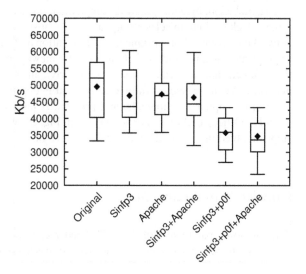

Considering the same scenarios, we have also tested the overhead introduced by the kernel modules on large data transfers by uploading a 500 MB file to an FTP server. Most packets will not be altered, but the conditions of the if statements in the kernel modules need to be evaluated, thus adding some overhead, which will eventually affect the net transfer rate. As we can see from Fig. 20, the more conditions need to be evaluated, the larger the effect on performance is.

Attacker Perspective

In the last set of experiments, we evaluated our approach from the point of view of an attacker trying to determine the OS of a remote host or the type of services running on it. In order to test how our approach can deceive attackers using Nessus, we audited the system with and without the deceptive kernel module enabled. Table 2 shows the results of the respective Nessus scans. The original system is a fully patched Ubuntu 12.04 server and has no known vulnerabilities. When no deception is used, the system is correctly identified, and all the information derived by Nessus is accurate. Next, we deceived both OS and service fingerprinting by exposing a Windows 7/Vista OS fingerprint and an Apache 2.2.1 service fingerprint. When the deception mechanism is enabled, the OS is misidentified accordingly, and the deceiving service fingerprinting leads to false positives in the identification of vulnerabilities.

Table 3 reports the results of scans performed with different fingerprinting tools. Clearly, our approach is able to effectively deceive several tools. For instance, we are able to alter the perception of the target system, even when the attacker uses either nmap or Xprobe++, which adopts a different probing scheme.

In conclusion, by intelligently crafting responses to SinFP probes, it is possible to force attackers into misclassifying a remote host as any of a broad variety of networked assets. For instance, a conventional Linux-based server can be fingerprinted as a network switch, an ADSL gateway, or even a printer. Of course, these fingerprints will cause attackers to derive an inconsistent map of the target network. We have successfully created SinFP deceptions for different network monitoring appliances, firewalls, and printers. A partial SinFP output for the case of an HP Officejet 7200 Printer is reported in Fig. 21, whereas Fig. 22 illustrates the steps involved in forcing the attacker to believe that the target device is a printer.

Drawbacks and Limitations

Altering some parameters of the TCP header can affect connection performance, when legitimate users actively use the protocol based on the modified parameters. In

Table 2 Results of Nessus scans

	Without deception	With deception
Device type	General purpose 85%	General purpose 65%
OS	Ubuntu 12.04 85%	Windows Vista 65%
Info	13	15^8
Low	0	0
Medium	0	2 (100%: false positives)
High	0	2 (100%: false positives)
Critical	0	0

Table 3 OS fingerprinting deception

Tool deception	Sinfp3	p0f	nmap	Xprobe++
None	Linux 3.0.x–3.2.x (94%) Linux 2.4.x–2.6.x (73%)	Linux 3.x	Linux 2.6. x–3.x	Linux 2.4.19–28 (94%)
Windows Server 2008	Server 2008/Vista/7 (100%) FreeBSD 7.0–9.0 (73%)	Windows 7/8	Unknown	Linux 2.4.26 (78%)
FortiGate Firewall	FortiGate Firewall (100%)	Unknown	Unknown	Linux 2.4.23 (94%)
NetBSD 5.0.2	NetBSD 5.0.2 (98%)	Unknown	Unknown	Linux 2.4.21 (92%)
Windows Server 2008 (Partial)	Server 2008/Vista/7 (98%) FreeBSD 7.0–9.0 (73%)	Windows 7/8	Unknown	Linux 2.4.26 (81%)
Windows Server 2008 (Partial)	Server 2008/Vista/7 (88%)	Unknown	Unknown	Linux 2.4.14 (81%)

```
score: 100: Printer: HP Officejet 7200
score: 100: Printer: HP Officejet Pro L7600
score: 73: Appliance: APC AP9319
```

Fig. 21 SinFP output for a remote host discguHP Officejet 7200 Printer

Fig. 22 SinFP Printer Deception

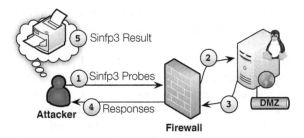

such cases, the proposed mechanism is not completely transparent, and drawbacks include those listed in the following.

Maximum Segment Size (MSS) This parameter defines the largest unit of data that can be received by the destination of the TCP segment. Modifying this value makes the host announce a different limit for its capabilities. Consider two hosts h_A and h_B, where h_B is the host being altered. Assume h_A sends a SYN packet with an MSS of 1460 and h_B responds with a SYN/ACK that has an MSS of 1480. Then, h_A will not send any segment larger than 1480 bytes to h_B, even if h_B may actually be able to handle larger segments. Note that h_A is not required to send segments of exactly 1480, but it is required not to exceed this limit. For the same reason, it is not possible to advertise a larger value than what hosts are actually able to handle.

Window and Windows Scale Factor These two parameters affect the TCP flow control, which regulates the amount of data a source can send before receiving an acknowledgment from the destination. A sliding window is used to make transmissions more efficient and control the flow so that the destination is not overwhelmed with data. The TCP window scale factor is used to scale the window size by a power of 2. The window size may vary during the data transfer while the scale factor is determined once at the time of establishing a connection. Modifying the window size can alter the throughput: if the window is smaller than the available bandwidth multiplied by the latency then the sender will send a full window of data and then sit and wait for the receiver to acknowledge the data. This results in lower performance.

Selective ACK Selective acknowledgment allows the sender to have a better idea of which segments are actually lost and which have arrived out of order. If we disable the SACK option for a host that supports it, we may limit performance, depending on the capabilities of the counterpart.

Conclusions

In this chapter, we presented a principled approach for manipulating outgoing traffic so as to induce an external view of the system that will deceive potential intruders into making incorrect inferences about important system characteristics, including OSs and active services. We demonstrated practical applicability of the proposed approach by presenting deception-based techniques for specifically defeating an attacker's effort to fingerprint OSs and services on the target system. Although experimental results show that our approach can efficiently and effectively deceive an attacker, some limitations still exist and more work remains to be done. In addition to some of the limitations listed in section "Drawbacks and Limitations", we need to consider that the proposed manipulation of outgoing traffic is limited to just some categories of traffic, and there might be other categories of traffic or other characteristics that attackers can use to infer the nature of the OS and services on a target host. Thus, our approach needs to be extended to address this scenario and make our solution more resilient and capable of defeating more sophisticated reconnaissance efforts.

References

1. P.K. Manadhata, J.M. Wing, An attack surface metric. IEEE Trans. Softw. Eng. **37**(3), 371–386 (2011)
2. Executive Office of the President, National Science and Technology Council, "Trustworthy cyberspace: Strategic plan for the federal cybersecurity research and development program," December 2011, URL http://www.whitehouse.gov/
3. M. Dunlop, S. Groat, R. Marchany, J. Tront, Implementing an IPv6 Moving Target Defense on a Live Network, in *Proceedings of the National Moving Target Research Symposium*, (Cyber-Physical Systems Virtual Organization, Annapolis, 2012)
4. Q. Duan, E. Al-Shaer, J.H. Jafarian, Efficient Random Route Mutation Considering Flow and Network Constraints, in *Proceedings of the 1st IEEE Conference on Communications and Network Security (IEEE CNS 2013)*, (IEEE, Washington, DC, 2013), pp. 260–268
5. J.H. Jafarian, E. Al-Shaer, Q. Duan, Openflow Random Host Mutation: Transparent Moving Target Defense Using Software Defined Networking, in *Proceedings of the 1st ACM Workshop on Hot Topics in Software Defined Networks (HotSDN 2012)*, (ACM, Helsinki, 2012), pp. 127–132
6. M. Albanese, A. De Benedictis, S. Jajodia, K. Sun, A Moving Target Defense Mechanism for MANETs Based on Identity Virtualization, in *Proceedings of the 1st IEEE Conference on Communications and Network Security (IEEE CNS 2013)*, (IEEE, Washington, DC, 2013), pp. 278–286
7. S. Jajodia, A.K. Ghosh, V. Swarup, C. Wang, X.S. Wang, *Moving Target Defense: Creating Asymmetric Uncertainty for Cyber Threats, vol. 54 of Advances in Information Security* (Springer, 2011)
8. V. Casola, A. De Benedictis, M. Albanese, Integration of Reusable Systems, in *A Multi-Layer Moving Target Defense Approach for Protecting Resource-Constrained Distributed Devices*, Advances in Intelligent and Soft Computing, (Springer, 2013)
9. V. Casola, A. De Benedictis, M. Albanese, A Moving Target Defense Approach For Protecting Resource-Constrained Distributed Devices, in *Proceedings of the 14th IEEE International Conference on Information Reuse and Integration (IEEE IRI 2013)*, (IEEE, San Francisco, August 2013), pp. 22–29
10. G.F. Lyon, *Nmap Network Scanning: The Official Nmap Project Guide to Network Discovery and Security Scanning* (Insecure, Sunnyvale, 2009)
11. D. Watson, M. Smart, G.R. Malan, F. Jahanian, Protocol scrubbing: Network security through transparent flow modification. IEEE/ACM Trans. Networking **12**(2), 261–273 (2004)
12. G. Shu, D. Lee, Network Protocol System Fingerprinting - A Formal Approach, in *Proceedings of the 25th IEEE International Conference on Computer Communications (INFOCOM 2006)*, (IEEE, Barcelona, Spain, 2006). https://doi.org/10.1109/INFOCOM.2006.157
13. D. Barroso Berrueta, A practical approach for defeating Nmap OS-Fingerprinting, January 2003., URL http://nmap.org/misc/defeat-nmap-osdetect.html
14. A. Rana, What is AMap and how does it fingerprint applications?, March 2014, URL http://www.sans.org/security-resources/idfaq/amap.php
15. C. Trowbridge, An overview of remote operating system fingerprinting, SANS Institute InfoSec Reading Room, July 2003.
16. F.H. Abbasi, R.J. Harris, G. Moretti, A. Haider, N. Anwar, Classification of Malicious Network Streams Using Honeynets, in *Proceedings of the IEEE Conference on Global Communications (GLOBECOM 2012)*, (IEEE, Anaheim, 2012), pp. 891–897
17. C.-M. Chen, S.-T. Cheng, R.-Y. Zeng, A proactive approach to intrusion detection and malware collection. Secur Commun Netw **6**(7), 844–853 (2013)
18. M. Albanese, E. Battista, S. Jajodia, V. Casola, Manipulating the Attacker's View of a System's Attack Surface, in *Proceedings of the 2nd IEEE Conference on Communications and Network Security (IEEE CNS 2014)*, (San Francisco, 2014), pp. 472–480

19. M. Albanese, E. Battista, S. Jajodia, A Deception Based Approach for Defeating Os and Service Fingerprinting, in *Proceedings of the 3rd IEEE Conference on Communications and Network Security (IEEE CNS 2015)*, (IEEE, Florence, 2015), pp. 253–261
20. P. Auffret, SinFP: unification of active and passive operating system fingerprinting. J. Comput. Virol. **6**(3), 197–205 (2010)
21. M. Zalewski, p0f v3 (version 3.06b), January 2012, URL http://lcamtuf.coredump.cx/p0f3/
22. R. Gula, Enhanced operating system identification with Nessus, February 2009, URL http://www.tenable.com/blog/enhanced-operating-system-identification-with-nessus

Next-Generation Architecture and Autonomous Cyber-Defense

Carol Smidts, Xiaoxu Diao, and Pavan Kumar Vaddi

Abstract This chapter introduces the motivation for and emerging developments in next-generation network architectures to enable autonomous cyber-defense (ACD), including promising studies on cyber-defense approaches and mechanisms applied to contemporary industrial control systems (ICSs).

Synopsis

This chapter introduces the motivation for and emerging developments in next-generation network architectures to enable autonomous cyber-defense (ACD), including promising studies on cyber-defense approaches and mechanisms applied to contemporary industrial control systems (ICSs).

Overview

Industrial control systems (ICSs) are playing crucial roles in contemporary industries, such as energy, transportation, chemical, and manufacturing. Researchers and engineers have studied and developed several large-scale cyber-enabled ICSs, such as Supervisory Control and Data Acquisition (SCADA) systems and Distributed Control Systems (DCSs). Generally, these ICSs are cyber-physical systems (CPSs) consisting of one or more control servers (CSs), several remote terminal units (RTUs), programmable logic controllers (PLCs), other intelligent electrical devices (IEDs), and sensors. These devices are usually task-oriented and lack any mechanisms of cyber-defense. Recently, to enhance productivity and lower the cost of industrial processes, an increasing number of stakeholders have sought to deeply

C. Smidts (✉) · X. Diao · P. K. Vaddi
Reliability and Risk Laboratory, Department of Mechanical and Aerospace Engineering,
The Ohio State University, Columbus, OH, USA
e-mail: smidts.1@osu.edu; diao.38@osu.edu; pawan.kumar@osumc.edu

© Springer Nature Switzerland AG 2019
C. Rieger et al. (eds.), *Industrial Control Systems Security and Resiliency*, Advances in Information Security 75, https://doi.org/10.1007/978-3-030-18214-4_10

integrate networking technology with control systems, which would allow them to effectively share business data and monitor control procedures with different departments in an enterprise or even with other business units or institutes. As ICSs are moving away from the traditional approach of isolated systems running proprietary control protocols on specialized software and hardware toward an increasing application of widely available, low-cost information technology (IT) networking systems, the threats and the consequences of cyberattacks have increased manifold. By utilizing networking technology to enhance productivity and lower the cost of data sharing for ICS, industrial organizations are exposing mission-critical ICSs to cyberthreats through exploitation of vulnerabilities in the connected CPSs. Breaking the natural isolation between cyberspaces and physical control processes may allow cyberattacks to interfere with normal productivity, destroy related equipment, or even lead to more catastrophic consequences to the environment and to humans.

Although various proactive and passive defense methods and technologies have been studied and practiced, such as firewall, diode transmission, Virtual Private Network (VPN), variant Intrusion Detection Systems (IDSs), and diverse Vulnerability Discovery Tools (VDTs), ICS networks are still vulnerable when facing intelligent and experienced adversaries with sufficient vulnerability information about the network architecture and components. Current cyber-defense applications can only perform limited actions, such as to warn the operators or bypass and reset the compromised devices, to resist attacks. These resistant actions may be noneffective or unnecessarily costly if the attacker acquired the knowledge of the defense behaviors. Furthermore, human factors in the process of cyber-defenses may deteriorate the correctness of detecting anomaly and selecting countermeasures due to the unreliability and limited knowledge and capacity of humans. For example, in a nuclear power plant, the overwhelming number of alarm signals that appear at control panels when an unexpected situation is detected may significantly affect and postpone the decisions made by plant operators. Sometimes, the delayed responses would leave enough time for the adversaries to reach their targets.

Hence, it is implicit that the next-generation networks for ICSs should be able to defend themselves against such threats, survive, and be available when required. Smith et al. [1] defined the ability of a network to defend against and maintain an acceptable level of service in the presence of challenges as network resilience for IT networks. Rieger et al. [2] expanded this idea of resilience to ICSs by further adding organizational resilience—"the ability of an organization to survive in the face of threats" to IT resilience—and defined that "a resilient control system is one that maintains state awareness and an acceptable level of operational normalcy in response to disturbances, including threats of an unexpected and malicious nature." With the ever-increasing threats to ICSs, it is therefore imperative that resilience be a necessary requirement of ICSs. Inspired by this theory, this chapter proposes an autonomous cyber-defense (ACD) approach based on resilient control design, which is a set of approaches and techniques enabling a cyber-physical system to defend from, detect, respond to, and recover from a series of cyberattacks without human intervention. Referring to the systematic architectural framework based on several major research initiatives such as ANSA, ATIS T1, and CMU-CERT, the ACD framework is based on D^2R^2, the real-time part of the resilience strategy [3] and our research as part of a Nuclear Energy University Program (NEUP) project, "Support for Reactor Operators in Case of Cyber-Security Threats" [4]. This strategy contains the seven steps that make up the real-time cycle known as "the reactive cycle," (1) defend, (2) detect, (3) evaluate, (4) predict, (5) analyze, (6) respond, and (7) recover, which will assist a system in rapidly adapting to challenges and attacks and maintain an acceptable level of service.

Implementing the ACD approach requires the existing cyber-systems of ICSs to evolve to a next-generation architecture, which can automatically select and activate the optimized strategies of cyber-defense to mitigate the risk of system functionality loss and minimize the severity of functional failures. This means that the new architecture should effectively be aware of hazards, efficiently predict the objective of threats, and reconfigure itself to minimize the hazard. Also, the architecture should be able to timely notify the system operators of the hazard and provide reasonable solutions or recommendations if the upcoming hazardous situation falls beyond the handling capabilities of the architecture itself. Finally, the new architecture should be capable to manage and maintain a knowledge database that records the experiences of prior defenses and to evolve itself for a better resistance to future threats. Based on this, in this chapter, we propose a next-generation architecture with ACD at its center. This architecture consists of the following three parts. (1) knowledge base, (2) autonomous cyber-defense, and (3) refinement. The refinement part of the architecture is based on the background cycle *DR* presented in Sterbenz et al. [3] and encompasses the activities diagnosed (i.e., root cause analysis, a cost-benefit analysis, and refine), which enable longer-term evolution of the system to enhance the approaches to the activities of the reactive cycle. The ACD performs its actions based on the challenge models and policies in the knowledge base [1], which is updated frequently through both human intervention and feedbacks from refinement. This architecture for network resilience is depicted in Fig. 1.

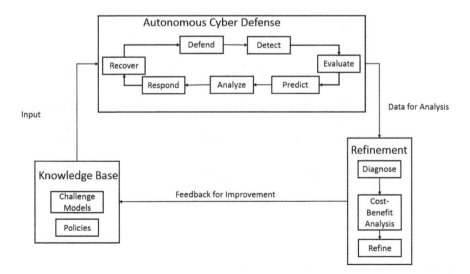

Fig. 1 Next-generation architecture

In the following sections, current practices in ICS cybersecurity, along with the proposed resilience strategy [3, 4], a resilience control design framework [1, 4], and an architecture, are discussed to establish a next-generation architecture for ICSs.

Understanding the Challenges

This section provides an overview of the architectures and implementations of current ICS cyber-networks. In addition, the challenges and countermeasures are introduced to lead to the requirements for the next-generation network architecture.

ICS Networks

The ICS network is a hierarchical, multilevel, real-time network with various cyber-physical components [5]. As shown in Fig. 2, a general ICS network architecture consists of several field sites, one or more control center, and maybe some remote access points. Components in the ICS network are deployed into various security levels. Data transmission between layers is restricted and inspected. For example, in a nuclear power plant, data can only be transferred from a high-security level to a

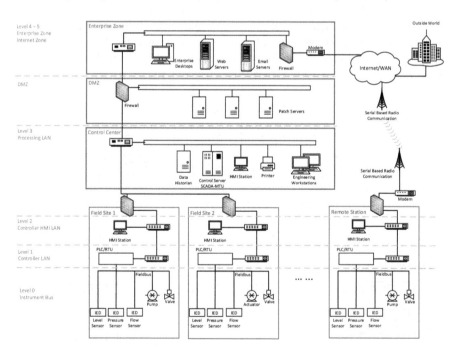

Fig. 2 Example architecture of ICS network of a power plant

lower-security level. The reverse is prohibited to guarantee that the operation derived from low-level networks will not impact the ones in the high-level networks. In the cases where bidirectional data transmission is required, a demilitarized zone (DMZ) will be used to play an intermediate role of data transfer between high-security level and low-security level networks.

Field sites (see the lower portion of Fig. 2) are functional points that execute real-time tasks, such as closed loop control or signal sampling. A field site usually encompasses one or more PLC/RTU for executing control process, IEDs/sensors and actuators for interacting with physical components, and a fieldbus connecting the involved cyber components. Examples include the controller area network (CAN) bus [6] in a vehicle or robot, the ARINC 664 [7] in an aircraft, and Profibus [8] or ControlNet [9] in power plants.

The fieldbus is time-critical, since the latency of data transmission between nodes will probably lead to serious consequences, such as the damage to facilities. Also, a fieldbus provides limited routing capability. Traditional fieldbuses use point-to-point connection, such as the RS-232 bus, since the interactions between these nodes are simple and non-frequent. The contemporary fieldbus enhances the connectivity between the controlled nodes, hence increasing the routing capability (i.e., the CAN bus that supports a maximum of 127 to 255 online nodes). In addition, fieldbuses are highly fault-tolerant. A residual bus usually exists as a backup for keeping connectivity.

The control center (see the middle section of Fig. 2) is a supervisory level accounting for monitoring and configuring the set points of control algorithms, as well as adjusting related parameters in the controller. At this level, human-machine interfaces (HMIs), such as digital instruments and displays, are utilized to provide process status information to the plant operators. Data services can store and replay the historical information of control processes. Also, there will be engineering stations performing diagnostics and prognostics tasks to prevent, identify, and recover from abnormal operations or failures.

The connections between control center and field sites are implemented by a Local Control Network (LCN) in a DCS, since the field sites are located at the same location as the control center. For the SCADA system, this connection is usually supported by a wide area network (WAN), such as the Internet or special wireless communications. An LCN is also a time-critical network since it will transfer critical control set point data, which will deeply impact the function of field sites and probably cause damage to the facilities at field sites. The defense of LCN is challenging due to the diversity of topology and the complexity of network protocols used in the LCN.

The enterprise zone (see the top portion of Fig. 2) is a business network performing general information tasks in an enterprise or power plant, such as financial planning or human resource management. Typically, the enterprise zone can access the Internet or other types of WAN.

Many contemporary ICSs can accept a connection from a remote station through a WAN (see the right side of Fig. 2). A remote station is an autonomous site that can implement real-time tasks, which have to be geographically located at a different location than the control center. As depicted in Fig. 2, the remote station may have its own fieldbus and local networks.

Due to the diversity and complexity of threats, the security strategies deployed into the ICS network cannot perfectly avoid threats. As opposed to general IT systems, existing ICSs (usually implemented by CPSs) are weak and vulnerable to possible cyberattacks. Several characteristics of the ICS reflect this problem:

- ICSs are task-critical. It is often difficult or impossible to interrupt the operation of ICSs for system updates or defense software installation. In a hard real-time system, a millisecond's latency will significantly interrupt the normal control process and cause abnormal consequences, such as unexpected high temperature and the damage of control components. Typically, ICSs are the most critical parts of these systems.
- ICSs are resource-constrained. Commonly, real-time control is implemented by embedded systems, such as PLCs and IEDs. This means that the control and data-acquisition application is running on a tailored, dedicated hardware platform. In some extreme cases, this application directly manipulates the hardware and peripherals without the use of an operating system. Therefore, many general cyber-defense techniques for IT systems, such as antivirus software, cannot be applied to the terminals belonging to an ICS. Additionally, the limitation of useful computing resources inhibits the deployment of monitoring and diagnostic facilities into the network, which makes the collection of cyberattack evidence difficult.
- Legacy systems exist in ICSs. Many well-aged systems in use have vulnerabilities that have not been discovered yet. These vulnerabilities are more likely to be exploited and leveraged by attackers because redesign of these systems is not affordable or feasible.
- ICSs contain commercial off-the-shelf (COTS) components. To lower cost, industrial organizations would like to use COTS components to replace dedicated functional components. For example, they would like to use real-time operating systems from the open-source community, instead of developing a dedicated task scheduler with corresponding device drivers from scratch. Also, they would probably seek a single-board computer (SBC) with a general-architecture central processing unit (CPU) (e.g., x86) rather than designing and developing a unique integrated circuit since the latter is more expensive and harder to maintain. However, COTS components introduce common vulnerabilities into ICSs. These common vulnerabilities increase the risk of ICS security since they are well known in the cyberattack community and are easily exploited by attack tools.

Challenges to ICS Networks

Threats to an ICS network consist of several aspects, such as the attack activity, the vulnerabilities of infrastructures and configurations, and other risk factors. This section will introduce and categorize these aspects and the corresponding defense methods that are widely used in current ICS networks.

Network-based ICSs are safety-critical systems that leverage network systems to deliver and share critical data to control the behaviors of physical systems. According to the classical fault-error-failure propagation path, faults, such as hardware flaws or software defects, will cause an erroneous state of a system component. Then, the abnormal state will activate other component's faults or abnormal states, named as fault propagation. Consequently, the anomalies will lead to a failure of system functionality. The fault propagation path is depicted in Fig. 3a–c.

Besides a classical fault propagation path, Fig. 3 also depicts the possible impacts from cyberattacks. As marked from (a) to (c), cyberattacks can affect the fault propagation path from the following three aspects:

- *Attack 1*: Cyberattacks can activate a dormant fault in an ICS. To launch such an attack type, the adversaries need to acquire sufficient information about the target system so that the faulty component can be located and accessed. A zero-day attack is an example of this type.
- *Attack 2*: Attacks can directly change the state of a component from nominal to erroneous even if no defect exists in such a component. For example, a Distributed Denial of Service (DDoS) attack can block the data transmission between the controllers and halt the control process.

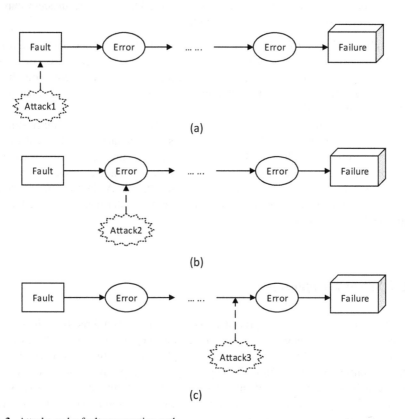

Fig. 3 Attacks and a fault propagation path

- *Attack 3*: Attackers can create or modify the propagation path that leads to a system failure on purpose. A straightforward example is that an attacker establishes a link to bypass the firewall in a network.

According to the empirical data collected over the past two decades [10], the primary threats to ICSs include:

- *Malware.* Several destructive computer viruses and worms have harmed files and hard drives, including the Melissa Macro Virus, the Explore.Zip worm, the CIH (Chernobyl) Virus, Nimda, Code Red, Slammer, and Blaster.
- *Social Engineering Scams.* Several attackers use cyber-tools as part of their information-gathering and espionage activities, such as phishing schemes in an attempt to steal identities or information for monetary gain.
- *Botnets.* Instead of breaking into systems for the challenge or bragging rights, bot-network operators take over multiple systems to coordinate attacks and to distribute phishing schemes, spam, and malware attacks.
- *Insider Threats.* Insiders may not need a great deal of knowledge about computer intrusions because their knowledge of a target system often allows them to gain unrestricted access to cause damage to the system or to steal system data. The insider threat also includes outsourcing vendors, as well as employees, who either intentionally or accidentally introduce malware into systems. Insiders may be employees, contractors, or business partners.

Another challenge to cyber-defense is the complexity of vulnerabilities existing in cyber-based systems. Exploited vulnerabilities are major weaknesses of a well-defended system [11]. More specifically, Table 1 categorizes common threats against ICS networks and their countermeasures based on domains (i.e., hardware, software, and network).

Because networks found in ICSs have varied characteristics, their attack surfaces differ [12–14]. In the past, threats and attacks were treated as individual events, which led to the individual application of countermeasures. However, researchers have proved that many cyberattacks are closely dependent and are usually conducted simultaneously in an attack scenario. As a result, the ACD framework should provide a synthetic consideration of threats and countermeasures to achieve a sensible defense strategy.

ICS Network Defenses

The types of cyber-defense approaches to be used are linked directly to the types of cyberattacks. Traditionally, these approaches are categorized into proactive defense and passive defense. Proactive defenses are applied prior to the occurrence of actual attacks, minimizing the probability of cyber-system compromise. On the other hand, passive defenses are activated during the threat development phase to resist the attacks and mitigate the risk of system failures. In this section, we briefly introduce current proactive and passive defenses.

Table 1 Threats and countermeasures in ICS networks

Network types	Domains		
	Hardware	Software	Network
Fieldbus	*Vulnerabilities and threats:* Hardware Trojan Illegal clones Side channel attacks (i.e., snooping hardware signals) *Countermeasures:* Tamper-resistant hardware (e.g., TPM) Trusted computing base (TCB) Hardware watermarking Hardware obfuscation	*Vulnerabilities and threats:* Software programming bugs (e.g., memory management, user input validation, race conditions, user access privileges, etc.) Software design bugs Deployment errors *Countermeasures:* Secure coding practice (e.g., type checking, runtime error, program transformation, etc.) Code obfuscation Secure design and development Formal methods	*Vulnerabilities and threats:* Networking protocol attacks Network monitoring and sniffing *Countermeasures:* Firewall Communication encryption
Local control network	*Vulnerabilities and threats:* Defects of switch or router *Countermeasures:* Patches and updates of firmware	*Vulnerabilities and threats:* Unauthorized access attacks Man-in-the-middle attack *Countermeasures:* Strong password Restricted operational control	*Vulnerabilities and threats:* Network monitoring and sniffing Broadcast or flood attacks *Countermeasures:* Firewall Intrusion prevention and detection
Remote access network	*Vulnerabilities and threats:* Utilization of hardware defects (Meltdown, Spectre) *Countermeasures:* Patches and updates of firmware	*Vulnerabilities and threats:* Database and SQL data injection Remote desktop control *Countermeasures:* Firewall Patches and updates of software	*Vulnerabilities and threats:* Social engineering attacks DDoS *Countermeasures:* Firewall VPN

Proactive defenses encompass several technologies and activities that decrease the probability of success of cyberattacks. Common proactive defenses include the encryption of message transmission, antivirus/malware software, vulnerability and penetration testing tools, etc. More specifically, the following are widely used cyber-defense technologies:

1. *Network Segregation:*

   - *A firewall* is a device or system that controls network traffic. A firewall can block communications, enforce secure authentication, and record information flow.

- *DMZs* are physical and logical subnetworks acting as intermediaries for connected security devices so that they can avoid exposure to a larger and untrusted network.
- *Virtual LANs* divide physical networks into smaller logical networks consisting of a single broadcast domain that isolates traffic from other VLANS.
- *Diodes*, or unidirectional network devices, are network appliances or devices that allow data to travel in only one direction [15].

2. *Host Security:*

- *Vulnerability assessment and penetration testing tools*, a set of approaches to detect any hardware and software bugs, known as "vulnerabilities," and assess the severity and impact levels of threats that can be caused by exploiting such vulnerabilities by finding an open door to penetrate the target system [16].
- *Antivirus/anti-malware software*, a program used to protect a computer system from being affected by viruses or other types of malware [17].
- *Virtual machines (VMs)*, a software-based application used to encapsulate the network applications and share the hardware resources between the encapsulated applications. Moving target defense (MTD) leverages VMs as a loadable kernel on servers to create a deception network.

3. *Access Control:*

- *VPN*, an encrypted channel for transferring data through networks.
- *Floating password*, a mechanism that generates passwords by using a predefined algorithm.
- *Authentication and authorization*, operators in an ICS environment are required to use multilevel authentication for accessing the systems. Physical keys are also used for such authentication, in addition to several layers of passwords and security questions.

Passive defenses are mainly structural defenses, which provide diverse redundant paths for connectivity. In addition, techniques such as trust boundaries are also used. Simultaneous links and redundant nodes are provided to route the signals around the failure and maintain the level of service required. Passive defense usually includes technologies such as Security Monitoring, IDSs, Intrusion Prevention Systems (IPSs), Security Audit Logging, and Security Incident and Event Monitoring (SIEM) systems.

Requirements for a Next-Generation Architecture

The National Institute of Standards and Technology (NIST) formulated several documents containing recommendations and guidance for assessing and avoiding cybersecurity issues in ICS networks [18–20]. These documents provide guidelines to protect the availability, integrity, and confidentiality of information systems in ICSs. These guidelines deliver basic requirements for the ACD and the next-generation architecture. According to the standards created for industrial systems, the basic requirements of cyber-systems in ICSs are as follows: [11, 21]

- *Safety.* An ICS network should always ensure the safety of humans, the environment, and equipment during its operation.
- *Availability.* In ICSs, rebooting may not be acceptable. Such outages must be planned and scheduled. There is a requirement of high availability and quick response to humans and other systems implementing emergency measures.
- *Integrity.* Software changes must be thoroughly tested and deployed incrementally throughout a system.
- *Confidentiality.* Encryption should be applied to data transmissions. Authentication and authorization should be deployed for critical data.

A comparison between general IT systems and ICS networks has been made [18]. Here, we extend this comparison to next-generation ICS networks, as shown in Table 2.

Defense-in-depth is a defense strategy that uses two or more different overlapping mechanisms or technologies to minimize the impact of security failures in ICSs. This strategy should be inherited by the ACD and the next-generation architecture. The end goal is to reduce the opportunities for an adversary to take advantage of the ability to move laterally through an entity's networks/systems and thereby to increase the cost of intrusion.

According to NIST's recommendations [18], an effective defense-in-depth architecture strategy for the survivability of the system in the face of an adverse challenge should include:

- The efficient utilization of firewalls to deny all unnecessary data transmissions
- The deployment of a DMZ to isolate different security levels of networks and filter dangerous data packages
- Backup facilities to avoid accidental outages of critical services
- A well-designed monitoring and diagnosis mechanism to evaluate hazards and risks at runtime

In addition to the proactive defense mechanisms of the abovementioned defense-in-depth strategy, the next-generation ICSs should be able to implement further proposed reactive defense mechanisms autonomously through a systematic process of context awareness and risk evaluation and deliver an acceptable level of performance in the face of adverse events [2]. In summary, the next-generation ICS networks should be resilient in nature.

Table 2 Differences between general information systems and ICSs

Categories	Current ICS network	Next-generation ICS network
Security focus	Protecting functional processes (e.g., control process, manufacturing process)	Protecting functional processes Recovery from abnormal statuses
Performance	Time-critical Modest throughput is acceptable	Time-critical Modest throughput is acceptable Recovery time is defined
Availability	Outages must be planned and scheduled in advance. High availability required, may necessitate redundant systems	Outages are not acceptable High availability required Predefined responses and activities for resilience are necessary
Risk management	Human safety is paramount, followed by protection of the process Fault tolerance is essential	Human safety is paramount, followed by protection of the process Fault tolerance and recovery are essential.
System operation	Hard real-time operating system is generally used, often without security capabilities built in Software changes must be carefully made	Hard real-time operating system is generally used Fault detection and situation awareness mechanisms are deployed Software and network can be automatically set to an adaptive configuration
Communications	Proprietary and standard communication protocols Networks are complex including several dedicated wire and wireless media	Proprietary and standard communication protocols Networks are complex including several dedicated wire and wireless media
Integrity	Software changes must be thoroughly tested and deployed incrementally throughout a system to ensure that the integrity of the control system is maintained in the ICS	Software changes must be thoroughly tested and deployed incrementally throughout a system to ensure that the integrity of the control system is maintained in the ICS
Accessibility	Components are usually isolated and require extensive physical effort to gain access to them	Components are well protected. The access to them will change according to the process and situation
Resource constraints	Resources are limited and tailored No extra resources for additional security solutions	Specific resources are prepared for system dynamic reconfiguration
Challenge tolerance	Employ redundancy and diversity for fault tolerance capabilities	Need to be attack tolerant in addition to being fault tolerant

Theory of ACD and Next-Generation Architecture

Strategy for Resilience

Given that we have established the requirement of resilience in next-generation ICS networks, in this section, we propose a strategy for resilience based on the $D^2R^2 + DR$ strategy [3] and our work [4]. This strategy consists of two phases.

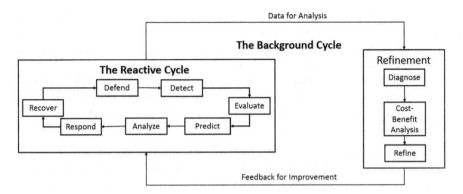

Fig. 4 The proposed resilience strategy

The first phase is a reactive cycle (i.e., the ACD that operates in real time to maintain the short-term resilience of the ICS network). The second phase is the background cycle (i.e., refinement that runs in the background to improve the reactive cycle's performance over time for long-term resilience). This strategy for resilience is depicted in Fig. 4.

The Reactive Cycle

The reactive cycle of this resilience strategy comprises of a cycle with the following seven steps, which are performed in real time and are directly involved in ICS network operation.

Defend The first step toward achieving a resilient ICS network is to defend the network against challenges and threats to normal operation. A set of defenses that are put in place as part of challenge tolerance, to reduce the probability of intentional design faults or non-intentional faults leading to a failure of the system and reduce the impact of a hostile attack, is the essence of this step. As threats evolve in time, these proactive defenses may be overcome by those challenges. Hence, the next step of the resilience strategy is to detect any such challenges that can penetrate these defenses.

Detect The second step in this resilience strategy is to identify the failure of defense systems and to detect challenges to the ICS network at both the system level and the individual component level. Whether the ICS network is challenged or not can be determined in three main ways as proposed by Sterbenz et al. [3] The first method is to detect anomalies in system operation based on understanding the normal behavior of components, both physical components (such as motors, pumps, etc.) and network components (such as the amount of traffic between nodes, data packet size, etc.) at given conditions and identifying any deviations from such expected behavior based on a set of predefined metrics. In order to do this, a complete understanding of the normal behavior of components of ICSs would be required. In general, it would be sufficient to have an understanding of normal behavior of components such as PLCs and IEDs that are susceptible to such attacks. When such an understanding of normal behavior of components is not available, one of the following two methods can be used to detect challenges. The first is to detect a deviation of ICS network components from desired functionality. This requires understanding the network service requirements at any given condition. The second is to detect errors, which can develop into total service failure at the level of components such as routers.

Evaluate Once an adverse event is detected, it is obvious that the next step is to evaluate the effects of the detected event or challenge on the ICS. Since ICS components are directly responsible for controlling the physical systems, an adverse event on an ICS network can lead to physical damage and, thus, financial loss on a large scale. The objective of this step is to evaluate the current state at both the system and the component level in terms of the resilience metrics based on the results from the detect step.

Predict If the detected challenge belongs to an existing attack category, then the available information will be used to identify the projected progression of the adverse event. If the detected challenge is an active attack that does not match with an existing attack category, it is important to predict the future steps of the attack to devise and implement an effective response strategy. This prediction is based on understanding the possible objective and profiles of the attackers [4].

Analysis It is a common practice in safety-critical industries to develop event trees and fault trees as part of a Probabilistic Risk Assessment (PRA) model to map the progress of small failures into large-scale disasters [23]. Once the results from the evaluation and prediction step are available, PRA models will be used to map the challenge to event trees and fault trees. This information is then used in a game-theoretic framework to generate response strategies [4]. The risk value (i.e., the probability the challenge evolves into a large-scale disaster) is calculated, and the response strategy involving the least amount of risk to the system is subsequently selected.

Respond The generated response strategies are implemented to effectively counter the challenge and protect the system from further damage. If the need arises, several remediation measures are activated to move the system into a degraded state of operation, such as disconnecting several components to alleviate the damage, where it operates with less than normal levels of performance.

Recover At the end of the response step (i.e., after the adverse event is over), the system may be in a degraded state. After detecting the end of the adverse event, the system should return to its normal state of operation. Generally, financial implications play a larger role in determining when to begin recovery of the system.

The Background Cycle

The second phase of this resilience strategy is the background cycle, which consists of two background operations that observe, analyze, understand, and learn from the reactive cycle to improve its performance in the long run.

Diagnose The first step of the background cycle is an offline process of root cause analysis to identify any non-intentional faults that may have crept into the system, which resulted in the adverse event. It is important to note that generally, faults are detected only when those cause observable errors. The objective of this step is to identify the root causes that resulted in a challenge that crept past the defenses and to identify any possible solutions to prevent the occurrence of similar events in the future. These solutions can range from improving the proactive defenses at a network level to complete overhaul of the structure.

Cost-Benefit Analysis Once the possible solutions are identified, a cost-benefit analysis needs to be performed to decide upon implementing the solutions that are simultaneously effective and have the least possible financial and performance implications.

Refine The final step of this strategy is to refine the behavior based on the past reactive cycles. Past real-time reactive cycles are analyzed to improve future performance in order to achieve higher efficiency in dealing with challenges. In addition, the new defensive measures identified through the abovementioned diagnosis and cost-benefit analysis steps can be added to tackle the newly identified faults. To perform this step, it is imperative that the system is able to evolve over time. Flexibility for such evolution at both the physical and network level should be considered while designing the system.

Resilient Control Design

Based on the abovementioned resilience strategy and the resilient control design proposed by Smith et al. [1] in this section, we present a feedback control loop-based approach toward achieving resilience in ICS networks. Figure 5 depicts a systematic approach for network resilience based on the resilience strategy in the form of a feedback control loop, called the resilience control loop [1]. In a traditional feedback control loop, a controller modifies the input given to the system under control based on feedback received by comparing the actual output to the expected output, with the objective of steering the system toward the desired state. The resilient control design is based on the same principle, in which the resilience manager, upon evaluating the resilience of the system using the data from challenge detection, guides the appropriate response mechanisms toward achieving the resilience target.

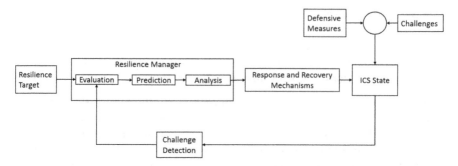

Fig. 5 Resilient control design–resilience control loop

The components of this control loop framework are described below:

- The reference level of ICS network resilience that is to be achieved is the resilience target given to the resilience control loop. This reference value is described using a set of resilience metrics such as availability, reliability, and performance level. This resilience target is a requirement set by the users, in this case the process control engineers and the network engineers in the industry.
- The second component of this resilience control loop constitutes the defensive measures put in place to defend the ICS network against any adverse events on the ICS network, and therefore the industrial systems, and maintain its ability to reach the set resilience target. A detailed study of all the possible challenges and their countermeasures is to be conducted in order to identify the best defensive measures to be put in place while making sure that those measures will not affect the regular speed and functionality required by the ICS networks. These defensive measures include passive measures, such as redundancy and diversity of components along with active measures like firewalls and antivirus software.
- Some challenges may overcome the defensive measures in place and cause the ICS to deviate from the reference resilience level. For example, a malware in a controller that exploits a zero-day fault or an active attack that exploits protocol vulnerability can overcome the defensive measures, or there may be a design fault, which can cause malfunctioning. Such observable errors are detected by the challenge detection component and characterized appropriately using various information sources.
- The resilience manager has three systems.

  1. An evaluation system that determines in real time whether the target resilience is achieved or not based on the outputs from the challenge detection and the input resilience target.
  2. A prediction system that predicts in real time the next steps in an active attack based on the existing knowledge base or the information about the attacker's possible objectives and profiles [4]. The prediction system utilizes PRA models, such as fault trees, to identify the final target of the attack and the corresponding steps to achieve it.

3. An analysis system that maps the predictions on a PRA model to identify the best possible response strategies using game-theoretic analysis.

   The resilience manager thus acts as a controller in a traditional feedback control loop by taking outputs from challenge detection and determining the best possible response actions to take in order to steer the system toward the resilience target.

• Response and recovery mechanisms implement the response strategies chosen by the resilience manager to steer the system toward normalcy in operation.

Resilience Components and Metrics

In order to implement the resilient control design, it is imperative to first understand how to define network resilience and the appropriate metrics to do so. As mentioned in the Guide to Industrial Control Systems (ICS) Security [22], the most important performance requirement of ICS networks in contrast to IT networks is a real-time response, which is a measure of quality of service. In ICS networks, the primary concerns in risk management are human safety, fault tolerance, loss of equipment, and loss of intellectual property. In addition, availability is also at most importance because control systems cannot be simply rebooted or repaired without planning days in advance to prevent affecting the production. Apart from that, ICS components, such as PLCs, are directly responsible for controlling the physical processes. So integrity of the data along with the interaction of ICS networks with the physical world must be considered.

Challenge tolerance and trustworthiness were presented by Sterbenz et al. [3] as the two disciplines that form the basis for network resilience in IT networks. The challenge tolerance discipline encompasses aspects of network resilience that are related to design and engineering, such as survivability, traffic tolerance, and disruption tolerance, whereas the trustworthiness discipline includes components that are measurable during operation, such as dependability, security, and performability. In contrast to IT networks, ICS networks are directly involved in physical processes. So the effects of problems in ICS networks, such as accidents, which can lead to injuries to operating personnel and damage to equipment, need to be considered while adapting the abovementioned resilience disciplines of IT networks for ICS networks. The two resilience disciplines and their components are presented in Fig. 6.

Challenge Tolerance

The first step toward building systems that can withstand challenges is to incorporate certain tolerance mechanisms into the system during the design stage itself. The challenge tolerance discipline encompasses such components of a network

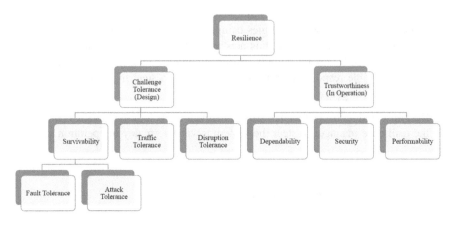

Fig. 6 Resilient disciplines

resilience, which deal with designing the network to tolerate challenges that can hinder desirable performance:

1. *Survivability:* Survivability is the ability of a network to survive in the presence of targeted attacks and multiple faults. So survivability is the combination of fault tolerance in the presence of single or multiple correlated and uncorrelated random faults and attack tolerance in the presence of targeted attacks by capable adversaries. Survivability is quantified using either set theoretic- and state machine-based formulation [24] or multidimensional Markov chains [25]. As mentioned above, fault tolerance capabilities and attack-tolerant abilities are highly essential in ICS networks. It is general practice in industry to provide redundant physical components as a part of fault tolerance. This practice should be extended to network components and networking routes to create survivable networks. Also, diversity of components is given importance to prevent common cause failures. In addition, defense measures such as firewalls and antivirus software should be incorporated.

2. *Traffic Tolerance:* This is the ability of a network to function in the presence of an unpredictably large legitimate traffic or an intentional injection of traffic as part of a DDoS attack. The possibility of large legitimate traffic that hinders performance is very low because ICS components are designed with sufficient computational power to respond in real-time while performing the required tasks along with designated channels to handle traffic from each node. Similar to survivability, traffic tolerance is an important aspect to consider while designing ICS networks. Measures such as frequent monitoring of traffic between nodes to identify sudden increase in traffic should be put in place. Usually, industrial components, such as controllers, are programmed to maintain last valid values for some time when experiencing delays in communication.

3. *Disruption Tolerance:* The ability of a network to communicate during disruptions, such as natural disasters, intentional physical attacks, and weak

connectivity, is disruption tolerance. Industries generally have physical security measures to prevent intentional damage to components. Nuclear power plants are built to withstand earthquakes and airplane crashes. Also, as mentioned above, controllers are programmed to maintain the last valid value to tackle the problem of occasional weak connectivity.

Trustworthiness

Once the required measures are put in place as part of challenge tolerance during the design phase, a set of measurable properties must be defined to monitor the system and verify its level of performance over time. Avižienis et al. [26] and Sterbenz et al. [3] defined trustworthiness as the "assurance that the system will function as expected with respect to measurable properties." The components of trustworthiness that measure the performance level include:

1. *System Dependability.* The ability of a system to deliver a service that can justifiably be trusted. When applied to ICSs, dependability is the measurement of confidence that an ICS can normally control the physical systems to satisfy the requirements of functionality, reliability, and safety.

   System dependability encompasses several important system attributes: reliability, availability, safety, integrity, and maintainability:

   System Reliability. The ability of a system to provide a continuous supply to its customers. ICS reliability is the ability of an ICS to continuously control a physical system to implement predefined tasks.

   System Availability. Availability is a measure of the degree to which a system is in an operable state and can be committed at the start of a mission when the mission is called for at an unknown (random) point in time. Availability as measured by the user is a function of how often failures occur and corrective maintenance is required, how often preventive maintenance is performed, how quickly indicated failures can be isolated and repaired, how quickly preventive maintenance tasks can be performed, and how long logistics support delays contribute to downtime.

   Quantifiable real-time reliability and availability measures can be used to assess the dependability and state of the system, when challenges are detected:

$$f(t)dt = \lambda(t)dt * R(t)$$

where $f(t)dt$ = the probability of failure in dt about t,

   $\lambda(t)dt$ = the probability of failure in dt about t, given that the system survived to time t

   $R(t)$ = the probability that the system did not fail prior to time t [23].

   Bayesian belief updating can be used to update the probability values in the presence of new evidence to get a real-time measure of the system resilience. In addition, measurements such as Mean Time to Failure (MTTF), Mean Time between Failures (MTBF), and Mean Time to Repair (MTTR) can be used as long-term measures of resilience to improve the performance of the reactive cycle over time.

System Safety. The ability of a system to prevent itself from endangering human life, health, property, or the environment. In the field of ICSs, system safety reflects the ability of an ICS to prevent dangerous events that cause hazards to humans, assets, or the environment.

System Maintainability. Maintainability is the ability of a system to be retained in, or restored to, a specified condition when maintenance is performed by personnel having specified skill levels, using prescribed procedures and resources, at each prescribed level of maintenance and repair.

System Integrity. ICS components, such as PLCs, take real-time decisions that affect the state of the system based on measurements from sensors, and those control-related decisions are in turn transmitted to physical components to steer the system toward the required state. Hence, it is implicit that the integrity of the data transmitted in an ICS during operation is of critical importance. This integrity can be evaluated in real time by comparing the signals with previously collected data.

When designing a networked ICS, we need to consider not only the dependability of physical systems and control applications but also the security related to cyber-systems, as shown in Fig. 7.

2. *System Security.* This is the contribution of a system to protect information so that unauthorized persons or systems cannot read or modify it while at the same time those who are authorized are not denied access to it.

   System security consists of system availability, integrity, and confidentiality:

   System Confidentiality. The ability of a system to preserve authorized restrictions on information access and disclosure, including means for protecting personal privacy and proprietary information.

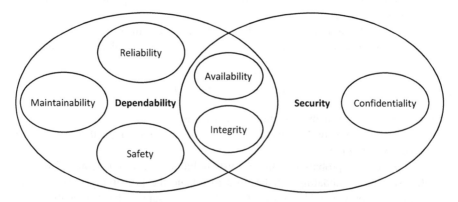

Fig. 7 Relation between dependability and security

Table 3 Real-time and long-term metrics of resilience

Real-time metrics	Long-term metrics
1. Reliability, availability (R(t), $\lambda(t)dt$)	1. MTTF, MTTR, MTBF
2. Integrity	2. Failure identification time [27]
3. Performance degradation	3. Recovery time [27]
	4. Performance loss over time [27]
	5. Total loss over time [27]
	6. Overall potential critical loss [27]

3. *Performability.* Sterbenz et al. [3] defined performability as "the property of a system such that it delivers performance required by the service specification, as described by QoS (quality of service) measures." Wei et al. [27] defined metrics such as performance degradation, performance loss over time, and total loss that can be based on operational and financial aspects of performance to quantify the performability aspect of resilience.

- Performance degradation due to an incident $\left(P_i^d\right)$ is defined as the difference between the Original System Performance (P_o) and the performance due to the incident (P_i). This is a real-time measure of performance that can be used to evaluate the resilience of the system.

$$P_i^d = P_o - P_i$$

Table 3 provides a list of real-time and long-term metrics of resilience. We refer the readers to Wei et al. [27] for a more detailed description of long-term metrics of resilience in ICS.

Implementation

In this section, we present a next-generation architecture based on the abovementioned resilience strategy, resilient control design, and the dynamic adaptation architecture proposed by Smith et al. [1], which realizes the resilient control design to attain network resilience in ICSs. It is implicit from the above section that such an architecture consists of subsystems, such as defense systems, challenge detection systems, evaluation, prediction and analysis systems, and response systems, in addition to the traditional communication systems of ICS networks. The resilience manager controls these subsystems based on information in a knowledge base, which contains extensively analyzed challenge models along with a set of decision-making policies. The next-generation architecture is depicted in Fig. 8 and its components are discussed below.

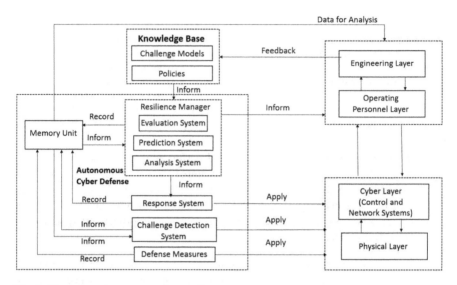

Fig. 8 The proposed next-generation architecture

Memory Unit

At the center of this architecture is the memory unit, a storage of historical data of the system, known challenges, and their outcomes (i.e., the behavior of the system and components during challenges), which transmits information between the various subsystems that constitute the reactive cycle. The memory unit records information from challenge detection components about detection and information about evaluation, prediction, and response steps from the resilience manager in real time. The memory unit also provides the challenge detection components any additional information required to determine the nature of the challenge. The memory unit also plays a major role in the refinement part of the architecture by storing data during real-time cycles for future analysis.

Resilience Knowledge Base

The resilience knowledge base contains two components. The first is the database of all possible challenge models created through extensive observation, analysis, and understanding of various threats to the ICS network. It is important to note that the memory unit is a storage of the behavior of the system during normal operation and adverse events, whereas the knowledge base contains challenge models that include the actions to be taken to defend the system in the face of challenges. Given that creating such a knowledge database requires a significant amount of financial investment in addition to human efforts, care must be taken in identifying the

appropriate threats. The threats can be classified based on both impact and probability of occurrence for appropriate allocation of resources. For example, more
resources need to be allocated toward developing models of threats with high
probability and high impact than the ones with low probability and low impact.
PRA techniques, such as event tree analysis and fault tree analysis, can be used in
evaluating the risk and identifying the probability of the detected error developing
into various large-scale disasters as a part of developing the challenge models [4].

The second component of the resilience knowledge base is a set of policies put in
place to facilitate easy decision-making by the resilience manager. For example, in
nuclear power plants, the reactor is shut down even if maintenance of a single
component of one auxiliary feed water system is delayed beyond the allotted time.
This decision is based on the policies present in place to prevent large-scale disasters.
In a similar way, policies need to be developed to determine appropriate response
and recovery mechanisms based on the acceptable levels of risk in ICS networks. In
addition, given that the nature and complexities of challenges change over time, the
challenge models and response strategies need to be updated periodically to adapt to
the latest threats. Policies need to be put in place for purposes such as determining
the time period of the updates, the scheduling of downtime for maintenance of
industrial equipment, and events of similar nature that require human intervention.

Defensive Measures: Defend Subsystem

As the first step of the resilient strategy discussed above, defensive measures are
required to be put in place to defend the network against challenges with minimum
possible impact. These defensive measures are primarily proactive in nature as they
are set in place in anticipation of known types of challenges. Smith et al. [1]
proposed that as challenges may vary broadly from simple link failures at the
topology level to complex malware, defensive measures must be applied across all
levels of the network. These include specific measures such as customized antivirus
software for diverse components across the ICS network to general measures such as
physical security in the industrial environment. As mentioned in the previous
sections, these measures can be permanent structural measures based on redundancy
and diversity and isolation through DMZs and data diodes, along with flexible
software measures such as antivirus software and firewalls. In addition, steps such
as updating firmware after careful testing and checking portable devices for malware
both before and after use are also part of the defensive measures requiring human
intervention. In ideal cases, these defensive measures should be able to completely
prevent any challenge from affecting the system. However, in realistic situations, it
would be sufficient if the challenges are contained at infected locations by measures
such as firewalls isolating the infected component and rerouting across the infected
nodes through a redundant path without affecting overall performance. The performance of defensive measures is recorded in the memory unit for future analysis to

improve the overall performance of the ACD system. A few defensive measures that need to be implemented in the next-generation ICS network architectures include:

- *Cyber situation awareness*, which is a technology that enables the cyber-system to be aware of the occurrence, feasible countermeasures, and possible consequence of cyberattacks by using the data sampled from network transitions. Cyber situation awareness consists of three important steps: "perception," "comprehension," and "projection" [28].
- *Moving target defense (MTD)* is a way to increase the difficulty for an attacker to exploit a vulnerable system by changing aspects of that system to present attackers with a varying attack surface [29]. Attack strategies include circumvention attacks, deputy attacks, brute force and entropy reduction attacks, probing attacks, and incremental attacks. Diversity defenses includes address space randomization, instruction set randomization, and data randomization.
- *AI algorithms* are used to solve the decision problems in cyber-defense [30]. Neural nets, DoS detection, computer worm detection, spam detection, zombie detection, malware classification and forensic investigation, expert systems, security planning, and intelligent agents against DDoS, search algorithms, learning algorithms, constraint solving, hybrid dynamical system [31], and adaptive reinforcement learning [32] are examples.

Figure 9 depicts the impacts of cyber-defensive measures on dependability and security attributes. The black points in the figure express that applying such a defense mechanism (given along the X axis) will probably increase or enhance the corresponding system attributes (given along the Y axis). In contrast, the white points express that an applied defense mechanism may decrease or degrade a corresponding attribute. For example, a firewall and a DMZ can isolate safety-critical networks from open networks (i.e., the Internet). By applying strict access rules, these techniques can effectively increase the confidentiality and safety of ICSs.

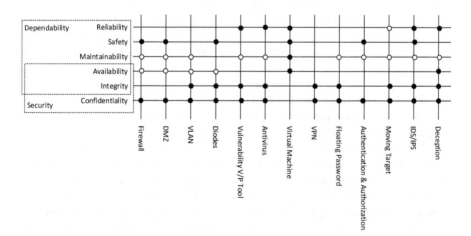

Fig. 9 Cyber-defense impacts on system attributes

However, these rules will probably increase the difficulty of system maintenance and decrease the connectivity and availability of some devices and components.

VLANs and diodes can logically separate subnets by configuring package routing at switches and routers. The routing configuration technically increases the difficulty of data access and enhances confidentiality and integrity. However, the increased complexity of network structures will negatively affect the maintainability and availability of ICS components.

Host security technologies, such as vulnerability verification tools and antivirus software, can detect the defects or malwares in an ICS's host and decrease the probability of a component's failure and improve the system's reliability. However, installing extra software will also increase the complexity of the ICS and decrease maintainability. If the extra software contains defects, reliability may ultimately decrease.

Modern defense technologies, such as MTDs, IPSs, and deception, increase the difficulty of conquering a target system so that they can effectively improve the confidentiality and integrity of the ICSs. In addition, MTDs will probably decrease the reliability and availability of ICSs because they periodically change the system configuration (e.g., access password) and may affect normal data access and system operations.

Challenge Detection Subsystem

The second step in the reactive cycle of the resilience strategy is to detect any challenges that the network faces, which can lead to a loss of required functionality. Smith et al. [1] proposed an incremental approach to challenge analysis, because with an increase in the availability of inputs from a wide range of information sources, perception about the nature of the challenge evolves. Lightweight detection mechanisms that do not consume significant computational power can always be active at local nodes, acting as the first level of defense, for prompt detection of anomalies and initiation of remediation mechanisms. Additional information can be accessed from the memory unit to determine the nature of the challenge, when the lightweight detection mechanisms fail to do so. Simultaneously, heavier mechanisms can be invoked incrementally only at affected nodes to better understand the nature of harder challenges. Implementing such an incremental approach enables the system for better allocation of resources in determining the nature of challenges.

Cyberattack detection strategies [33] are divided into (1) IDSs, (2) misuse detection/misbehavior detection, (3) signature-based approaches, and (4) anomaly detection.

The traditional solution against cyberattacks is the deployment of IDSs. IDSs recognize and alert awareness regarding the presumed occurrence of attacks by identifying the features of packages being transferred through the network. Technical solutions of IDSs can be (1) Embedded Programming Approach, (2) Agent-Based Approach, [34] (3) Software Engineering Approach, (4) Artificial Intelligence

Approach, (5) Cyberattack Detection in Cloud, and (6) Cloud Intrusion Detection Service Requirements.

Machine Learning [35] and Big Data [36] are novel trends of implementing intrusion detection. Promising studies in this field include:

- Artificial Neural Networks
- Association Rules and Fuzzy Association Rules
- Bayesian Networks
- Clustering
- Decision Trees
- Ensemble Learning
- Evolutionary Computation
- Hidden Markov Models
- Inductive Learning
- Naïve Bayes
- Sequential Pattern Mining
- Support Vector Machine

However, IDSs can only detect known attacks accurately. Traditional intrusion detection algorithms employ features extracted from historical attack data to identify an attack signature in real time or detect the anomalous behavior of the system [37]. They cannot detect unseen attacks. Also, the specification used by an IDS to detect attacks is deeply dependent on human security experts to analyze historical attack data [33].

An active attack or malware disguised as a component failure can lead to incorrect response measures, which can further increase damage to the system. So it is essential to identify the nature of the challenge. A comprehensive dataset with features from all existing types of cyberattacks and safety events needs to be created. The observed challenge can then be mapped onto the developed feature vectors to correctly estimate the nature of the challenge. Figure 10 shows the steps in detecting the nature of the detected event K [4].

The detection mechanisms make use of information on both safety events and cyber-events to effectively determine the nature of the challenge. So the detection mechanisms should be implemented to identify challenge propagation at the following four levels:

- *Cyber to Cyber:* Mechanisms such as packet data trace and traffic between nodes can be used to identify the propagation of errors to identify the nature of the challenge.
- *Cyber to Physical:* Physical measurements of inputs received by physical components from cyber components can be compared with historical data or a mathematical model of the process to identify or detect the nature and location of the fault.
- *Physical to Physical:* A faulty bearing or gear can increase the load on other components and can affect the performance of a physical system. Even though this type of fault propagation is not the primary focus of ICS cybersecurity, such

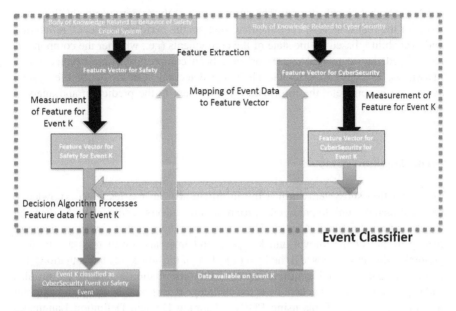

Fig. 10 Challenge—nature detection

detection measures should be implemented for correct identification of the nature of the challenge.

- *Physical to Cyber:* A faulty sensor can send incorrect readings to a controller, which can in turn lead to erroneous inputs to other physical components in the system. This problem becomes more prevalent with the increasing application of smart sensors in ICSs. Smart sensors and wireless sensors can be compromised to report incorrect data to the corresponding controllers. Diversity in the types of sensors can result in the quick detection of such errors.

Resilience Manager

Once the challenge detection subsystems identify a problem, alerts are sent to the memory unit. The network resilience manager then analyzes this context data and, based on the policies and challenge models from the resilience knowledge base, takes appropriate response measures. In order to do so, the resilience manager uses the following subsystems.

Evaluation Subsystems

The function of the evaluation subsystem is to evaluate the current state of the system in terms of the given resilience metrics based on the information from the challenge detection subsystems. The current state of the system can be evaluated in terms of the real-time resilience metrics, such as the performance level, reliability, and availability, based on the state of the components (i.e., whether the components have failed due to natural causes or have been compromised due to attacks and malwares), as determined by the challenge detection systems. Once the system evaluation is complete, the information is then sent to the prediction and analysis subsystems.

Prediction Subsystems

The prediction subsystems utilize the information from the resilience knowledge base and memory unit to predict the progress of the adverse event when the detected event is of a recognized category. However, when the detected challenge does not belong to any of the known attack types, fault trees and event trees are used in predicting the future steps and the final objective of the attack. In such situations, the prediction systems need to utilize any available information regarding the profiles and capabilities of the attackers from the challenge detection subsystem and external information sources. Tools using PDDL (Planning Domain Definition Language) can be utilized to generate attack paths to predict the future steps of the attacker in the ongoing adverse event [38].

Analysis Subsystem

The predicted steps of the adverse event are mapped onto a PRA model to identify the progression into a potential event that can cause large-scale damage to the system. It is general practice to develop PRA models for potential accidents due to component failures in safety-critical industries. These PRA models should be modified to include the failures of electronic components due to malware and attacks. Once the adverse event progression is identified based on mapping the prediction results onto the PRA models, an effective response strategy is developed using game-theoretic analysis. The attack on the ICS is modeled as a noncooperative game [4], in which the risks and rewards are calculated using the PRA models. The analysis system utilizes the information from challenge models and policies in the resilience knowledge base. A response strategy with least possible safety, performance, and financial impacts is then selected and the corresponding response mechanisms are activated.

The resilience manager informs the operating personnel about the evaluation and response strategies. The operating personnel can then decide to continue the ACD process or manually intervene. The resilience manager also records all of the information generated during the evaluation, prediction, and analysis steps in the memory unit for future analysis.

Response Subsystem

The response strategies generated by the resilience manager are implemented to effectively counter the challenge and defend the system from further damage. The response subsystem can contain mechanisms such as standby servers to take over operation, dynamic routing capabilities to isolate the infected nodes, and adaptive firewalls that can effectively defend the system based on the new information from the resilience manager. For example, once the attack is mapped to a fault tree in the analysis phase, the path the attacker is most likely to take to move upward in the fault tree is identified and the components in that path can be disconnected from the compromised component as a part of the response [4]. In simple cases, the system can continue functioning at normal performance levels. However, based on the level of risk involved, the response subsystem also needs to activate remediation mechanisms, such as completely disconnecting components for the system to transition into a state of degraded performance until the attack is over.

Refinement

The data that is stored in the memory unit during a real-time cycle is used for improving the performance of ACD in future cycles. The first step is to perform diagnosis to identify the root cause of the challenge. As mentioned above, a fault can be detected only when it develops into an observable error. Through the diagnosis, once the root cause of a challenge that went past the defense measures in the past cycle is identified, the propagation of the fault can be mapped and the challenge models can be updated for future cycles. However, this refinement and diagnosis need to be performed manually. Further research should consider automating this process.

Conclusion

This chapter proposed a set of fundamental theories and technologies to implement the resilient control design as part of ACD in a next-generation architecture. Ongoing research is directed toward implementing the proposed architecture in real case studies [4].

Appendix 1: Acronyms

ACD	Autonomous Cyber-Defense
AI	Artificial Intelligence
ANSA	Advanced Networked System Architecture
ARINC	Aeronautical Radio, Incorporated
ATIS	Alliance for Telecommunications Industry Solutions
CAN	Control Area Network
CERT	Cyber Emergency Response Team
CMU	Carnegie Mellon University
CPS	Cyber-Physical System
CPU	Central Processing Unit
CS	Control Server
DCS	Distributed Control System
DDoS	Distributed Denial of Service
DMZ	Demilitarized Zone
HMI	Human-Machine Interface
ICS	Industrial Control System
IDS	Intrusion Detection System
IED	Intelligent Electrical Device
IT	Information Technology
LAN	Local Area Network
LCN	Local Control Network
MTD	Moving Target Defense
MTTF	Mean Time to Failure
MTTR	Mean Time to Repair
NIST	National Institute of Standards and Technology
PDDL	Planning Domain Definition Language
PLC	Programmable Logic Controller
RTU	Remote Terminal Unit
SBC	Single-Board Computer
SCADA	Supervisory Control and Data Acquisition
TCB	Trusted Computing Base
TPM	Trusted Platform Module
VDT	Vulnerability Discovery Tool
VPN	Virtual Private Network
WAN	Wide Area Network

References

1. P. Smith et al., Network resilience: A systematic approach. IEEE Commun. Mag. **49**(7), 88–97 (2011)
2. C.G. Rieger, D.I. Gertman, M.A. McQueen, Resilient control systems: Next generation design research, in *Proceedings of the 2nd Conference on Human Systems Interactions (HSI)*, (2009), pp. 632–636
3. J.P.G. Sterbenz et al., Resilience and survivability in communication networks: Strategies, principles, and survey of disciplines. Comput. Netw. **54**(8), 1245–1265 (2010)
4. C. Smidts et al., Support for reactor operators in case of cyber-security threats, in *ANS Annual Winter Meetings 2017*, vol. 117, (2017), pp. 929–932

5. K. Stouffer, J. Falco, K. Kent, Guide to Industrial Control Systems (ICS) Security recommendations of the National Institute of Standards and Technology, in *NIST SP-800-82*, (2008)
6. M. Farsi, K. Ratcliff, M. Barbosa, An overview of controller area network. Comp. Contr. Eng. J. **281**(21), 113–120 (1999)
7. ARINC, Full duplex switched ethernet (AFDX) data bus. *ARINC 664*, 2007 (2015)
8. E. Tovar, F. Vasques, Real-time fieldbus communications using Profibus networks. IEEE Trans. Ind. Electron. **46**(6), 1241–1251 (1999)
9. B.F. Lian, J.R. Moyne, D.M. Tilbury, Ethernet, ControlNet, and DeviceNet. SIMULATION, 66–83 (2001)
10. Homeland Security, Recommended practice: Improving industrial control systems cybersecurity with defense-in-depth strategies. ICS-CERT, 1–56 (2016)
11. J. Jang-Jaccard, S. Nepal, A survey of emerging threats in cybersecurity. J. Comput. Syst. Sci. **80**(5), 973–993 (2014)
12. S. Hansman, R. Hunt, A taxonomy of network and computer attacks. Comp. Secur. **24**(1), 31–43 (2005)
13. M. Uma, G. Padmavathi, A survey on various cyber-attacks and their classification. Int. J. Network Sec. **15**(5), 390–396 (2013)
14. C.A. Myers, S.S. Powers, D.M. Faissol, Taxonomies of cyber-adversaries and attacks: A survey of incidents and approaches, in *Lawrence Livermore National Laboratory Report No. LLNL-TR-419041*, (2009). https://doi.org/10.2172/967712. URL https://e-reports-ext.llnl.gov/pdf/379498.pdf
15. H. Okhravi, F.T. Sheldon, Data diodes in support of trustworthy cyber infrastructure, in *Optimization and Security Challenges in Smart Power Grids (Energy Systems)*, ed. by V. Pappu, M. Carvalho, P. Pardalos, (2010), pp. 203–216. URL http://web.mit.edu/ha22286/www/papers/CSIIRW10.pdf
16. S. Shah, A modern approach to cybersecurity analysis using vulnerability assessment and penetration testing. Int. J. Elect. Commun. Comp. Eng. **4**(6), 47–52 (2013)
17. P. Vinod, V. Laxmi, M.S. Gaur, Survey on malware detection methods, in *Proceedings of the 3rd Hackers' Workshop on Computer and Internet Security*, (2009), pp. 74–79
18. National Institute of Standards and Technology, Security and Privacy Controls for Federal Information Systems and Organizations. NIST SP-800-53, Rev. 4 (2012)
19. National Institute of Standards and Technology, An Introduction to Information Security. NIST SP-800-12, Rev. 1, p. 85 (2017)
20. National Institute of Standards and Technology, Guide for Applying the Risk Management Framework to Federal Information Systems. NIST SP-800-37, Rev. 1, p. 93 (2010)
21. V. Gunes, S. Peter, T. Givargis, F. Vahid, A survey on concepts, applications, and challenges in cyber-physical systems. KSII Transact. Int. Inform. Syst. **8**(12), 4242–4268 (2014)
22. Stouffer, K., Pillitteri, V., Lightman, S., Abrams, M., and Hahn A., Guide to industrial control systems (ICS) Security. NIST SP-800-82, Rev. 2 (2015)
23. N.J. McCormick, *Reliability and Risk Analysis: Methods and Nuclear Power Applications* (Academic Press, New York, 1981)
24. J.C. Knight, E.A. Strunk, K.J. Sullivan, Towards a rigorous definition of information system survivability, in *Proceedings of DARPA Information Survivability Conference and Exposition (DISCEX 2003)*, vol. 1, (2003), pp. 78–89
25. P.E. Heegaard, K.S. Trivedi, Network survivability modeling. Comput. Netw. **53**(8), 1215–1234 (2009)
26. A. Avižienis, J.C. Laprie, B. Randell, C. Landwehr, Basic concepts and taxonomy of dependable and secure computing. IEEE Transact. Depend. Secure Comp. **1**(1), 11–33 (2004)
27. D. Wei, K. Ji, Resilient industrial control system (RICS): Concepts, formulation, metrics, and insights, in *Proceedings of the 3rd International Symposium on Resilient Control Systems (ISRCS 2010)*, (2010), pp. 15–22
28. U. Franke, J. Brynielsson, Cyber-situational awareness: A systematic review of the literature. Comput. Secur. **46**, 18–31 (Oct. 2014)

29. J. Li, J. Yackoski, N. Evancich, Moving target defense, in *Proceedings of the 2016 ACM Workshop on Moving Target Defense (MTD 2016)*, (2016), pp. 69–79
30. E. Tyugu, Artificial intelligence in cyber-defense, in *2011 3rd International Conference on Cyber Conflict*, (2011), pp. 1–11
31. R. Colbaugh, K. Glass, Proactive defense for evolving cyber-threats, in *Proceedings on the 2011 IEEE International Conference on Intelligence and Security Informatics (ISI 2011)*, (2011), pp. 125–130
32. M. Zhu, Z. Hu, P. Liu, Reinforcement learning algorithms for adaptive cyber-defense against Heartbleed, in *Proceedings of the First ACM Workshop on Moving Target Defense (MTD 2014)*, (2014), pp. 51–58
33. J. Raiyn, A survey of cyber-attack detection strategies. Int. J. Secur. Appl. **8**(1), 247–256 (2014)
34. F. Pasqualetti, S. Zampieri, F. Bullo, Attack detection and identification in cyber-physical systems. IEEE Trans. Autom. Control **58**(11), 2715–2729 (2013)
35. A.L. Buczak, E. Guven, A survey of data-mining and machine-learning methods for cybersecurity intrusion detection. IEEE Commun. Surveys Tutor. **18**(2), 1153–1176 (2015)
36. R. Zuech, T.M. Khoshgoftaar, R. Wald, Intrusion detection and big heterogeneous data: A survey. J. Big Data **2**(3) (2015). https://doi.org/10.1186/s40537-015-0013-4. URL: https://journalofbigdata.springeropen.com/track/pdf/10.1186/s40537-015-0013-4
37. R. Mitchell, I.-R. Chen, A survey of intrusion detection techniques for cyber-physical systems. ACM Comput. Surveys (CSUR) **46**(4), 1–29 (2014). https://doi.org/10.1145/2542049, art. 55
38. K. Tiwary, S. Weerawardhana, I. Ray, A. Howe, PDDLAssistant: A tool for assisting construction and maintenance of attack graphs using PDDL, in *Proceedings of the ACM Conference on Computer and Communications Security 2017 (CCS 2017)*, (2017). URL: https://pdfs.semanticscholar.org/9a02/7475fb90d342bf52c709e73a1351d5613481.pdf

Part IV
Human System Interface

Fault Understanding, Navigation, and Control Interface: A Visualization System for Cyber-Resilient Operations for Advanced Nuclear Power Plants

Christopher Poresky, Roger Lew, Thomas A. Ulrich, and Ronald L. Boring

Abstract Cyber is a buzzword these days—the destabilizing, invisible, insidious battleground that is shaking everything from financial institutions, hospitals, and airports to average individuals in an ever increasingly Internet-dependent world. The cybersecurity challenges facing nuclear power plant industrial control systems are much the same as for other industries where maintaining high availability of services is imperative. In this chapter, the authors discuss the development of cyber-resilient industrial control system technologies for nuclear power plants with a focus on operator support systems. These concepts culminate in the design and demonstration of the Fault Understanding Navigation and Control Interface (FUNCI) visualization system for cyber-resilient operation. Finally, the authors present the results of a study involving nuclear power plant operators.

Introduction

Cyber is a buzzword these days—the destabilizing, invisible, insidious battleground that is shaking everything from financial institutions, hospitals, and airports to average individuals in an ever increasingly Internet-dependent world. Because reporting on cyber-attacks stokes fears that each new event is unprecedented, unpredictable, and untraceable, it's easy to conclude that nothing is safe and we must expect that the worst can and will happen to every cyber asset in our society. Coupling this mentality with a critical pillar of the US energy infrastructure— nuclear energy—that endures the gold standard for scrutiny in both the public and

C. Poresky (✉)
Department of Nuclear Engineering, University of California–Berkeley, Berkeley, CA, USA
e-mail: chrisporesky@berkeley.edu

R. Lew
Department of Virtual Technology and Design, University of Idaho, Moscow, ID, USA
e-mail: rogerlew@uidaho.edu

T. A. Ulrich · R. L. Boring
Department of Human Factors, Idaho National Laboratory, Idaho Falls, ID, USA
e-mail: thomas.ulrich@inl.gov; ronald.boring@inl.gov

© Springer Nature Switzerland AG 2019
C. Rieger et al. (eds.), *Industrial Control Systems Security and Resiliency*, Advances in Information Security 75, https://doi.org/10.1007/978-3-030-18214-4_11

regulatory eye, sensational disaster fantasies can run wild [1, 2]. The reality is quite different.

While it can be tempting to think that cybersecurity concerns are commonly shared by many industries, it is still imperative to fully appreciate the nuance introduced by nuclear power's unique characteristics and how this nuance frames its unique position in the energy landscape. For example, currently operating nuclear power plants (NPPs) are almost entirely disconnected from the Internet. However, "cyber" is not synonymous with the "Internet." The terminology is used to distinguish from physical attacks or intrusions. Systems can be compromised without physical access, even if systems are disconnected from the Internet. We must then focus on the most "cyber"-relevant aspect of NPPs—their industrial control systems (ICSs).

The cybersecurity problem facing NPP ICSs is much the same as for other ICSs, where maintaining high availability of services is imperative: there are nonstop, active threats that can cause loss of service or even loss of assets due to damage [3–5]. Cyber-events can cause cascading feedback along the electrical grid with other electrical power generators [6, 7]. Furthermore, falling victim to a "hack" can result in global political repercussions [8, 9]. Fortunately, there are many best practice solutions that can and should be commonly applied to mitigate cybersecurity threats: designers can eliminate vulnerabilities through network infrastructure planning and access rules, they can maximize robustness of their plans through continuous policy reassessment and maintenance, and they must actively manage consequences at all network levels [10].

However, there are many things that set NPPs apart from other electric power sources. While their primary purpose is to deliver substantial amounts of clean, reliable electricity, they also have a very heavy safety and security focus. Compared to other electric generators, they have a unique consequence space including environmental contamination and adverse effects to public health in the form of radioactive particles, nuclear material theft and proliferation concerns, and a heightened profile as targets due to their sheer size, energy density, and the mere fact that they are nuclear facilities of high symbolic value [8, 11]. Even taking plants off-line for short periods of time could have cascading consequences for critical infrastructure and US economics.

In some ways, an examination of these unique cybersecurity concerns finds that NPP ICSs are already hardened; for example, their connectivity to the Internet is heavily restricted by US Nuclear Regulatory Commission (NRC) regulations. This intense cybersecurity protection posture is not without its caveats, though, because existing NPPs may cling to antiquated practices and technology in an effort to avoid a re-evaluation of their licensing basis. NPPs are modernizing control systems in a piecemeal fashion over several outage cycles to be able to continue cost-effective operation beyond their original 40-year licenses [12]. Through this process, NPPs have incrementally increased their capacity factors and efficiencies. Modifying nonnetwork-independent existing systems with new digital networked systems provides performance opportunities but could also introduce cyber-vulnerabilities [13, 14]. In contrast to existing NPPs, newly build NPPs that are still in the design phase have a significant opportunity to consider and incorporate cybersecurity from the ground up.

Development of New Systems for New Reactors

While there are many new NPPs being constructed around the world, the USA has not enjoyed the same fervor for new plant construction. As a result, the operational concepts and systems of NPPs in the USA were primarily developed decades ago and were heavily influenced by US naval reactor operations. In the interim since their construction, ICSs and operations have advanced in other less regulation-constrained domains such as oil and gas, chemical, pharmaceutical, and manufacturing process control. However, more than a handful of start-up companies in the USA are developing new NPP designs with a variety of advanced characteristics, such as being relatively small, operating at higher temperatures, being ready for flexible generation and cogeneration, being more fuel-efficient, and being passively safe with fewer moving parts [2]. The engineers at these companies make many of their design choices to incorporate lessons learned from the rich operating experience of NPPs in the USA and around the world to deliver NPPs that are more efficient, economical, reliable, safe, and suited to support our modern society. With these goals in mind, some companies have already begun to rethink the way they'll operate and control their NPPs. There are a variety of new technologies relevant to operations and monitoring that nonnuclear industries are adopting, such as machine learning, with the potential to reduce both investment and operating costs, improve plant performance, and augment existing safety and security practices. The nuclear industry has been hesitant to fully employ these technologies as it must navigate first-adopter regulatory scrutiny and accompanying uncertainty, financial concerns, and equipment compatibility issues. Adoption with existing plants is further complicated by the fact that they must also design and modernize systems while continuing to simultaneously operate the plants. With new nuclear designs, the opportunity to take cutting-edge developments and realize their full potential is ready to be seized. These technologies can fundamentally change the way designers and operators think about safety and security—not as separate burdens and constraints on performance—but as complements and assets to the operation of these advanced NPPs.

The central "new" technology pertinent to cybersecurity for new NPP designs is the digital control room. Digitalization brings the promises of added functionality, optimization, and flexibility. It also casts a shadow of doubt and uncertainty about the possible safety and security vulnerabilities brought along, especially with respect to cybersecurity. How can the promises be fulfilled and the doubts quashed? To start, plants must adopt best practices for securing ICSs, such as designing for defense-in-depth, segregating and air-gapping control communication buses, and engineering systems to fail safe even if digital control systems are compromised. Cyber-attack tools are widely available and easy to use; to illustrate the point, a simple Internet search for "low orbit ion cannon" shows how easily a prepackaged (albeit now commonly safeguarded against and benign) cyber-attack tool can be obtained. A high-target entity like a nuclear utility might contend with tens of thousands of cyber-attacks a year. Luckily, countermeasures exist to defend against common attacks, but events like Stuxnet have demonstrated that well-funded attacks by motivated agents are not always preventable [15]. The nature of cybersecurity is a game of cat and mouse—of measures and countermeasures. New plants are equipped with more sensors, computational power, and better algorithms than ever before. With thorough design studies and an iterative approach, the authors posit that digital systems can symbiotically serve to improve safety and security while simultaneously improving performance.

The advent of automation reduces the need for human intervention and allows the incorporation of dynamic, fault-tolerant techniques for maintaining system availability. When improperly implemented, automation can carry with it issues of relevance to cybersecurity, such as operator loss of situation awareness, unauthorized access to vital controls, and foreign manipulation of plant models used for system monitoring. To avoid these pitfalls, NPP designers must implement automation to capitalize on the strengths of algorithms and logic and must complement the critical-thinking and subjective reasoning skills of human operators. Through careful design of a hybrid human-machine system, automation can elevate the robustness and resiliency of the ICS to cybersecurity threats.

Cutting-edge sensor technologies are also changing the performance and cybersecurity landscapes. Wireless sensors significantly expand the set of viable locations and reduce the installation and maintenance costs compared to wired sensors requiring long cables, routing hardware, and even containment penetrations [16]. In doing so, they have the potential to unlock a richer world of data that can be used to inform automation and operator mental models of the plant's state and trajectory. Sensing innovations are not only hardware-based; there are also new benefits being realized in software and processing techniques for sensor data. Model-based sensors and virtual sensors incorporate physical understanding of the consistent field of data points and sampling locations, providing a continuous basis for online calibration and fault detection of sensors. Grounding sensor trustworthiness in a physical context is a key step toward reconstruction of faulty sensor values and resiliency in the face of instrumentation compromise. At the same time, these technologies enable a more judicious approach to sensor installation and may eschew the need for costly off-line calibration. The caveat with this set of technologies is a plant's increased dependence on robust, validated software and signal processing. The authors argue, however, that the maintenance issues for wireless sensors, such as bandwidth and remote access cybersecurity concerns, are far more manageable than for a cabled sensor that may have many yards of cable available for any number of physical faults.

New Operational Philosophy and Operator Support Systems

The impetus for a new philosophy of NPP operations is not only due to the allure of new technologies. Advanced nuclear companies have the desire and, in the USA, the need to prioritize flexible generation and/or cogeneration to maximize their contribution to grid integrity and deep decarbonization while enriching their designs' economic appeal. Dynamic power output and flow diversion associated with load-following and cogeneration represents a fundamental shift in day-to-day operations at NPPs in the USA. Transient operation will become normal operation, and understanding the interplay between myriad plant processes under shifting conditions will be vital to performance optimization. Managing this added level of complexity necessitates advanced operator support systems (OSSs) that draw from digitalization and human-oriented automation.

A well-designed OSS will present the operator with information in a way that supports an intuitive understanding of plant status and a conceptual model of event progression. It will draw on the data collected from modern sensor technology and online, real-time, automated calculations to provide a trustworthy picture of system conditions. One key opportunity afforded by digitalization is the capacity for task-based displays and contextual levels of detail. Through the distillation of relevant information, the OSS guides the operator to make quick, well-informed judgments. Judicious presentation of information improves focus and promotes evaluation of potentially meaningful relationships across plant processes. Finally, coupling this heightened awareness of system details with sophisticated monitoring and prognostics achieves the dual goal of performance improvement and cyber-defense.

Cybersecurity is a cross-domain concern, and effective policies must therefore be multifaceted. While it is true that information technology (IT) monitoring and support is crucial, ICS operators will play a key role in cybersecurity by preparing to deliver reliable, safe performance—even in the face of cyber-attacks. Operators may also be the first to identify cyber-maleficence. An OSS that establishes and continuously renews operator confidence in system states is a valuable tool alone, but coupling it with monitoring and prognostics enables the operator to be flexible and remain confident in fast-paced, unforeseen situations. What's more is that the hallmark of a truly powerful monitoring and prognostics system is its ability not only to proactively detect and isolate problems but to identify them as well. If operators acquire the information necessary to understand that an ICS has been infiltrated, the repercussions of the intrusion, and the remaining options for proceeding, they are vastly better equipped to handle the situation than if they were to simply make an emergency call to IT personnel and lose faith in the ability to operate the compromised system.

Demystifying cybersecurity for plant personnel is key to remaining agile and resilient in the face of system compromise. By designing an ICS and adhering to an operations philosophy that is resilient in the face of even the most unlikely of consequences, even from the cyber-realm, risk is dramatically decreased. Enabling rapid decision-making efforts like quarantining certain actuators and rerouting signals on-the-fly makes the difference between a system that suffers for its connectivity and one for which it prospers.

Fault Understanding, Navigation, and Control Interface (FUNCI)

With few currently operating NPPs transitioning to digital control and none having done so completely, development of the theorized OSS is an active field of research. Researchers at the Idaho National Laboratory (INL) Human System Simulation

Laboratory (HSSL) have been developing and demonstrating prototypical digital control systems for NPPs. A recent study involving a large US utility and NPP operators focused on the digitalization of the NPP turbine control system (TCS) [17]. While this digital system introduced several new modernization-related concepts, it also afforded the opportunity for something more experimental—the Fault Understanding, Navigation, and Control Interface (FUNCI, pronounced fun-key). FUNCI is a proof-of-concept prototype for an intelligent operator aid that could be integrated into a full-fledged OSS [18].

The TCS is an attractive application for FUNCI because it has real-world significance in the nuclear industry and beyond, its physics and operation are relatively simple, and it is a key candidate for innovation for load-following and cogeneration operations. Because the TCS relies on information exchange and physical feedback with the electricity grid, and therefore, other generators and consumers, understanding and managing its dynamic performance is of utmost importance for both safety and economics. Furthermore, the likelihood of adoption for such a system that uses a degree of automation is significantly higher for the TCS than for safety-related systems, such as the reactor protection system (RPS). The constraints on what information can be used and manipulated by FUNCI are therefore relaxed in comparison to more regulation-sensitive systems.

The essential function of FUNCI is a dedicated intelligent alarm system. One undesirable feature of existing NPPs is the potential for "alarm flooding," in which cascading alarms cause a large array of annunciator panels to flash and blare loud sounds. The operators must scan the annunciator panels before determining which alarms are most pressing to address and that are most representative of the issue. Even with more modern alarm lists, operators must scroll through hundreds or even thousands of alarms to determine the root cause. FUNCI aims to both minimize the number of alarms necessary to describe an issue and to provide information that can be used to solve problems before any other alarms are triggered. As such, FUNCI serves as an early warning system by using prognostics and context to evaluate system health and guide the operator toward a solution. Furthermore, alarms can be differentiated by levels, meaning that they can take the form of notifications that can sit in the background or of urgent messages that must be immediately addressed, with further granularity included as necessary. Finally, the alarms in FUNCI carry with them explanatory information that elucidates the cause and origin of the message. The main objectives of making alarms more intelligent are to minimize distracting and misleading false positives and extraneous information while simultaneously detecting issues at a much higher sensitivity and earlier time than is conventionally possible, which provides an operator with the ability to react deliberately and armed with good situation awareness. In some cases, alarms are made reductive by filtering nonessential information. In other cases, alarms are made additive by combining multiple alarms and even augmenting that alarm information with intelligent prognostics information.

A key aspect of an OSS is that it aids operators rather than supplanting them. This means that the design approach for the development of an OSS is not to list plant functions and allocate them between human and machine but rather to identify what tools the operator needs and then determine the technical feasibility of delivering those tools. This approach should also be applied when specifying the OSS features that contribute to ICS cyber-resiliency. NPP operators are not, nor are they expected to become, cybersecurity experts, although future operators will need a fundamental understanding of cybersecurity concepts and training to handle scenarios with cyber-complexities. This means that the appearance and functionality of human-machine interfaces (HMIs) should be realized through human-centered design for operators and not for IT engineers. One example of how this concept is practically applied is that useful metrics for attributing a system fault to cyber-attack should be communicated in clear terms rather than IT-specific technical jargon. If a temperature reading were somehow compromised by cyber-attack, the operator needs only to know that they should not trust that reading and that it has been compromised. From there, they can inform IT staff for the purpose of assessment and recovery, but can simply continue operating without trusting that measurement value. The lazy designer might display error codes or network ID information, but that would comprise extraneous information and reduce the utility of the OSS.

Form and Function of FUNCI

FUNCI, pictured below in Fig. 1, was designed to be a simple, unobtrusive, and clean stand-alone module that could be easily integrated into INL's existing proto-typical TCS display [19, 20]. We added it to a full-scope plant simulator with the entire array of traditional indications and alarms already available, so that it only added new functionality but would not override any existing sources of information. The idea was that it could be used if useful and ignored if not. The discoverability and intuitiveness of FUNCI was one of our primary interests in evaluating its design. It should be noted that FUNCI was created to be part of a display that is not interactive. That means FUNCI had no capacity for user input outside of actions that would affect its detection and conclusions about fault signatures.

In addition to the stationary box or "dock" for FUNCI, the system also flags indicators or components that are the subject of any current warning message. These indicators take the form of numbers in small red circles, as shown in Fig. 2, and were designed to be reminiscent of "notification bubbles" in many popular mobile operating systems and social media applications. While the design conventions of control rooms in currently operating NPPs share little in common with those of personal electronic devices that people use daily, the control rooms of tomorrow could draw on guidance from popular technology to ensure that they are intuitive and reduce the learning curve for operators to understand indications and alarms.

NPP digital control room interface designers must strike a balance between ease of use and maintenance of operator situation awareness. Following a dull screen display philosophy, FUNCI was designed to restrict the use of color to only those situations that demand rapid attention. By using a color-coded notification box and salient red indication and component flags, FUNCI swiftly guides the attention of

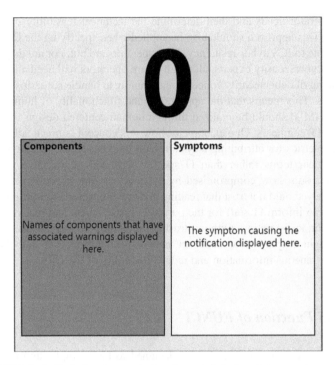

Fig. 1 FUNCI dock displaying the monitoring, but not fault detection state along with text to describe where components and symptoms information are displayed upon detection of a fault

operators to important information, without overwhelming them or saturating the display. When no faults are detected, the notification box remains blue, and the box itself consists of different shades of gray, in keeping with a dull screen style. No extraneous design elements are included in FUNCI, and the information provided is meant to be complete and optimally informative without being distracting or requiring too much concentration. Ideally, FUNCI can convey important messages that operators can fully comprehend with a quick glance before they move on to resume tasks or begin addressing issues.

With the move toward digital instrumentation and control, reinforcing operator trust in the system is of utmost importance. While it is true that analog instrumentation and control can also be faulty without operators realizing it (e.g., sensor calibration, physical degradation, connection issues), designers would be wise to tackle the parallel problems with digital systems from the outset. If operators are presented with guidance from digital systems that have no explanatory content for their solutions, they may not trust those systems. For this reason, FUNCI provides messages describing its warnings, as shown in Fig. 2, and offers supporting information so that the operator not only trusts the system but is able to communicate its findings after the fact and fully understand the plant state as well.

Because digital systems have the opportunity to leverage data-based approaches, such as online machine learning, it can be tempting to propose trained models for

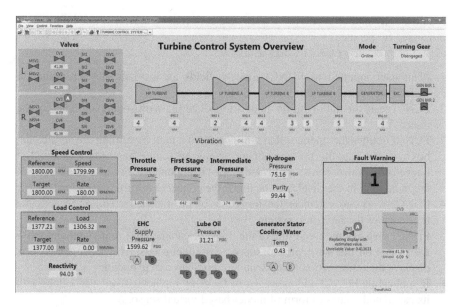

Fig. 2 FUNCI display embedded within a prototype turbine control system overview. The figure displays FUNCI in an active warning state in which a spoofed value for control valve #3 (CV3) was detected, and an alert was provided to the operator

prognostic purposes. However, "black box" models should not be extrapolated beyond the fundamental characteristics of their training dataset, do not provide physically meaningful relationships between inputs and outputs, and may artificially emphasize certain system features while minimizing others [21, 22]. Due to the lack of traceability, this represents an approach to automation that might prove difficult for facilitating operator trust. Even if operators were comfortable and effective with using these systems, they would be powerless to understand the underlying physics for faults and their corresponding solutions. In contrast, FUNCI is a physics-based fault detection and prognostics system. It uses physical relationship checks that represent mass, momentum, and energy balances to pinpoint inconsistencies and their sources. Using this information, FUNCI is also able to monitor trends and predict fault trajectories. The result is that the operator understands both the given situation and the basis for FUNCI's conclusions and recommendations.

A fundamental characteristic of FUNCI is that it does not automate any operation of the NPP or any decision-making. It leaves all control and practical problem-solving to the operators. In this way, FUNCI assists the operators by drawing on the strengths of computational systems and leaving the operators to draw on their unique cognitive reasoning and problem-solving abilities as human beings. The goal of a FUNCI-like system is not to automate the operator out of the control room but to provide the information necessary for the operator to act appropriately and optimally in all situations, whether those situations are known or not. FUNCI establishes a functional bridge between data collected by machines and information processed by people.

If a fault occurs, FUNCI works as follows:

1. Plant data inputted to FUNCI models yields an inconsistency according to FUNCI's physical models.
2. FUNCI tests the data with its models to isolate the fault source and identify the issue.
3. FUNCI notifies the operator that it has detected a fault by turning the notification box red and flagging the component where the source has been isolated. The flagged component is also replicated in the "Components" box as a visual copy of the indication in the display to facilitate the connection for the operator.
4. FUNCI works to calculate the trajectory of the faulty system, as well as the trajectory of the fault-free system.
5. FUNCI displays the trend line of the relevant parameter(s) in its fault and fault-free states on the same plot to support its conclusions. The "Symptoms" box displays this information along with any text-based explanation.
6. If applicable, FUNCI addresses sensor faults by replacing displayed values with its calculated values (a form of model-based virtual sensors).

We tested this method via a case study in which operators were given four different scenarios. Operators were not informed of the scenario details ahead of time, and we observed how they used the system and conducted a debrief interview after each scenario. The details of the case study are described in the next section.

Case Study Scenarios

Our case study was a subsection of scenarios within a larger study on a prototypical digital TCS. To substantiate the study findings, the study used a crew of three licensed reactor operators with over 20 years of combined operating experience at a US pressurized water reactor. The studies were also supervised by an experienced shift supervisor and operator instructor from another plant. The general format of each scenario in our study was to brief the operators that they had a task related to the TCS, start the scenarios, and then initiate scripted faults.

To test FUNCI and study its utility and discoverability in a general sense, we planned four separate scenarios that could be representative of cyber-attacks in a digital control system. These scenarios primarily rely on the compromise of digitally displayed information, or "spoofing" of indications. One other scenario represents a compromise of control input, where a value is entered "between" the interface and the actuation of the control. The scenarios are listed below.

Vibration Fault (Spoof)

During turbine ramp-up, a sudden increase in the turbine ramp rate to 80 MW/min was introduced, which represents the highest ramp rate that can occur before setting off an alarm, but is higher than required by any normal operation. FUNCI warns that the ramp rate is high and outside of normal operating parameters. The plant will trip if the operators do not take action.

Ramp Rate Surge (No Spoof)

During turbine ramp-up, a sudden increase in the turbine ramp rate to 80 MW/min was introduced, which represents the highest ramp rate that can occur before setting off an alarm, but is higher than required by any normal operation. FUNCI warns that the ramp rate is high and outside of normal operating parameters. The plant will trip if the operators do not take action.

CV Fault 1 (Spoof with Text)

A fault on control valve 3 was inserted to close the valve during full-power operation while also "spoofing" the value to match control valve 2 so that the change wasn't visibly noticeable on the turbine control system overview display. FUNCI detects the issue and warns that the display value does not match the estimated value that would be consistent with turbine pressure readings. After a delay, the system will replace the display value with its estimated value so that the operators can better understand the valve position.

CV Fault 2 (Spoof with Trend)

A fault on control valve 3 was inserted to close the valve during full-power operation while also "spoofing" the value to match control valve 2 so that the change wasn't visibly noticeable on the turbine control system overview display. FUNCI detects the issue and shows a warning that the valve display position has changed from what would be consistent with turbine pressure readings. It also shows a trend of estimated vs. displayed valve position over time. After a delay, the system will replace the display value with its estimated value so that the operators can better understand the valve position.

Case Study Results

In all four scenarios, the operators noticed that FUNCI displayed a warning and considered its messages. For the scenarios with spoofed indications, they did not act right away, but they made quick decisions once they were supported with information from FUNCI. There was some learning curve, with the operators learning to watch FUNCI more closely in successive scenarios. One shortcoming was that in the absence of audible indicators, the operators didn't notice a warning was indicated until they looked directly at FUNCI, leaving them to sometimes miss messages that were displayed in a time-sensitive fashion.

For the first scenario concerning the vibration fault, the operators were suspicious of the bearing warning before seeing FUNCI's assessment. They felt that their suspicions were confirmed by FUNCI and were able to make the decision to contact field personnel to check the turbine for corroborating signs of vibration. They also noted that instrumentation and control personnel would want to check the readings to determine if the sensor was bad.

For the second scenario, the operators noticed that the ramp rate surged, as a direct result of FUNCI's warning. They decided to trip the turbine, which is what would have happened as an automatic safety mechanism if they had taken no action, but they avoided the power level exceeding the approved technical specification safety threshold.

For the third scenario, the operators noticed FUNCI's message and still used their analog indications to confirm their mental model of the situation before feeling satisfied with the display. They didn't need to take any action because the three control valves that are not faulty were able to compensate for the faulted valve.

The fourth scenario was similar to the third except that the operators were able to trust FUNCI more readily. The two primary contributing factors to their reported greater trust were their familiarity with the system and the trend plot illustrating the spoofed value (in contrast to the text-only representation contained in the third scenario) and FUNCI's calculated value for valve position.

Lessons Learned

Overall, the operators responded very positively to FUNCI. The operators found it to be useful and a definite enhancement to the digital control system. They found that it assisted them in identifying the root cause of issues so that they could focus their decision on what mitigating actions to take. It should be noted that during the interviews, the operators emphasized that they felt anxious about the onus placed on them to quickly identify the root cause of issues during NPP operation. However, they found that FUNCI did provide them with early warning before their traditional alarms and monitoring practices would reveal issues. More specific to FUNCI as an augmentation of a digital control system, the operators explained that FUNCI assisted them in locating pertinent information during fault scenarios and that they found its visual implementation to be intuitive.

The operators also made a number of suggestions about considerations for future design iterations. They wished the notification messages in FUNCI would scroll and include time stamps so that they wouldn't miss any useful information and they would be able to track event progression. Based on the typical (and possible) number of alarms, careful management and presentation of this alarm list would be necessary to prevent it from becoming overwhelming or distracting. They also said that an audible component of the system would be helpful to better draw attention when they weren't actively looking at the system overview display. Trend plots seemed to be a better form of supporting evidence than text-based messages. The operators also discussed the potential usefulness of being able to call up and dismiss notifications on FUNCI in the same vein as acknowledging analog alarms. They said that a color code for differentiating root cause components from heavily affected and lightly affected components would also help them to prioritize their response to faults and anticipate system evolution. Finally, they seemed interested in having a dedicated screen or station for a system like FUNCI. Gathering these observations from actual end users and incorporating them into future design iterations of FUNCI is an example of formative evaluation [23]. It is up to the designer's best judgment which recommendations to take and how far to take them, but the simple reaction from the operators helps paint a picture of whether the system could be useful and how it would be used.

Conclusions and Discussion

While this initial study provided some useful insights and early positive feedback motivating the continued development of systems like FUNCI for operator support, there are still important further steps that must be taken. For one, it will become necessary to collect a body of quantitative data through the establishment of evaluation metrics and larger subject sample sizes in future studies to support any concrete conclusions about the value of FUNCI-like OSS applications. In addition to engaging more users as participants, we will add a greater range of functions and capabilities to FUNCI, enabling tests across a wider range of scenarios and other plant systems. While FUNCI's physics-based models were quite effective for the narrow set of scenarios in this case study, we will explore alternative prognostics engines and modeling methods in order to optimize fault detection, isolation, and identification methods for the larger possible set of plant faults. These future studies will largely be applied in a non-light water reactor (LWR) facility such as the Compact Integral Effects Test (CIET) facility at the University of California, Berkeley, which represents the primary systems in a prototypical fluoride salt-cooled, high-temperature reactor (FHR) [24]. Distinctions will therefore be made between the applicable aspects of FUNCI to currently operating LWRs and other prototypical reactor designs.

Recognizing the body of work remaining in the full development of FUNCI, there are a number of promising takeaways from this initial study. For one, the authors' assertion that cyber-related information can be used by operators constructively in a minimally demanding way was supported through the ease with which the operators were able to begin using FUNCI. It was integrated into existing plant diagnostic and prognostic information so that operators did not feel the burden of new training, but were able to immediately realize workload benefits. Furthermore, the case study demonstrated that visual representation of information relevant to cybersecurity was intuitive and gave them comfort in handling cyber-diagnostics, such as identifying faulty sensors and navigating compromised indications on digital plant displays. Finally, the operators' thoughts and recommendations in post-study interviews along with the data related to their performance are an invaluable part of the iterative design process that provides incremental validation along the way. Ultimately, this case study of FUNCI shows promise that cybersecurity is not a prohibitive concern in widespread adoption of powerful digital OSSs for NPPs and that human operators are capable of rapidly adapting to and mitigating new threats when their abilities are properly augmented by human-centered design.

Acknowledgments The lead author is a Ph.D. student in Professor Per Peterson's Thermal Hydraulics Laboratory in the Department of Nuclear Engineering at the University of California, Berkeley. The development work for this project was funded through a graduate fellowship provided by the US Department of Energy's Nuclear Energy University Program. Additional funding for carrying out the operator-in-the-loop study was provided by the US Department of Energy's Nuclear Energy Enabling Technology and Light Water Reactor Sustainability programs. This information was prepared as an account of work sponsored by an agency of the US government. Neither the US government nor any agency thereof, nor any of their employees, makes any warranty, expressed or implied, or assumes any legal liability or responsibility for the accuracy, completeness, or usefulness of any information, apparatus, product, or process disclosed or represents that its use would not infringe privately owned rights. References herein to any specific commercial product, process, or service by trade name, trademark, manufacturer, or otherwise does not necessarily constitute or imply its endorsement, recommendation, or favoring by the US government or any agency thereof. The views and opinions of authors expressed herein do not necessarily state or reflect those of the US government or any agency thereof.

References

1. A. Huhtala, P. Remes, Quantifying the social costs of nuclear energy: Perceived risk of accident at nuclear power plants. Energy Policy **105**, 320–331 (2017)
2. R.J. Budnitz, H.H. Rogner, A. Shihab-Eldin, Expansion of nuclear power technology to new countries–SMRs, safety culture issues, and the need for an improved international safety regime. Energy Policy **119**, 535–544 (2018)
3. J. Leclair, Cybersecurity and the nuclear environment, Nuclear News, 2016, pp. 30–31
4. B. Parisi, S. Wares, G. Bessis, P. Nicholson, Advanced cyber-attacks on global energy facilities, March 2014, Marsh Risk Management Research whitepaper, 2014. https://www.oliverwyman.com/content/dam/marsh/Documents/PDF/US-en/Advanced%20Cyber%20Attacks%20on%20Global%20Energy%20Facilities%20MRMR-03-2014.pdf

5. Nuclear Industry Summit 2016 Working Group, Working Group 1 Report: Managing Cyber Threats, March 2016, Washington, D.C., 2016. http://nis2016.org/wp-content/uploads/2016/02/Working-Group-1-Report-Managing-Cyber-Threats.pdf
6. World Institute for Nuclear Security, WINS International Best Practice Guide: Group 4 – Implementing Security Measures: 4.3 - Security of IT and IC Systems at Nuclear Facilities, March 2014, Vienna, 2014. https://wins.org/document/4-3-security-of-it-and-ic-systems-at-nuclear-facilities/
7. Lloyd's, Business blackout: The insurance implications of a cyber-attack on the U.S. power grid, Emerging Risk Report – 2015, July 2015. https://www.lloyds.com/~/media/files/news-and-insight/risk-insight/2015/business-blackout/business-blackout20150708.pdf
8. World Energy Council, World Energy Perspectives: The road to resilience | 2016, London, England, 2016. https://www.worldenergy.org/wp-content/uploads/2016/09/The-road-to-resilience_Financing-resilient-energy-infrastructure_Report.pdf
9. World Institute for Nuclear Security, WINS International Best Practice Guide: Group 3 – People in Nuclear Security: 3.1 - Developing Competency Frameworks for Personnel and Management with Accountabilities for Nuclear Security, July 2015, Vienna, 2015. https://wins.org/document/3-1-developing-competency-frameworks-for-personnel-and-managment-with-account abilities-for-nuclear-security/
10. World Institute for Nuclear Security, WINS International Best Practice Guide: Group 2 – Managing and Communicating Security Information: 2.3 - Information Security for Operators: Challenges and Opportunities, February 2014, Vienna, 2014. https://wins.org/document/2-3-information-security-for-operators-challenges-and-opportunities/
11. World Institute for Nuclear Security, WINS International Best Practice Guide: Group 4 – Implementing Security Measures: 4.11 - Effectively Integrating Physical and Cyber Security, May 2015, Vienna, 2015. https://wins.org/document/4-11-effectively-integrating-physical-and-cyber-security/
12. J. Naser, R. Fink, D. Hill, J. O'Hara, Human factors guidance for control room and digital human-system interface design and modification, November 2004, Electric Power Research Institute (EPRI) Technical Report (TR)-1008122, Palo Alto, 2004
13. International Atomic Energy Agency (IAEA), Computer security at nuclear facilities, December 2011, IAEA Nuclear Security Series Report No. 17, Vienna, 2011. https://www-pub.iaea.org/MTCD/Publications/PDF/Pub1527_web.pdf
14. C. Poresky, C. Andreades, J. Kendrick, P. Peterson, Cyber security in nuclear power plants: Insights for advanced nuclear technologies, September 2017, Department of Nuclear Engineering, University of California, Berkeley, Publication UCBTH-17-004, Berkeley, 2017. http://fhr.nuc.berkeley.edu/wp-content/uploads/2017/09/TH-Report-UCBTH-17-004.pdf
15. R. Langner, Stuxnet: Dissecting a cyberwarfare weapon. IEEE Secur. Priv. 9(3), 49–51 (2011)
16. J.B. Coble, P. Ramuhalli, L.J. Bond, J.W. Hines, B.R. Upadhyaya, Prognostics and health management in nuclear power plants: A review of technologies and applications, July 2012, Pacific Northwest National Laboratory (PNNL) Report No. PNNL-21515, Richland, 2012
17. R. Boring, T. Ulrich, R. Lew, C. Kovesdi, B. Rice, C. Poresky, Z. Spielman, K. Savchenko, Analog, digital, or enhanced human-system interfaces? Results of an operator-in-the-loop study on main control room modernization for a nuclear power plant, September 2017, Idaho National Laboratory (INL) Report No. INL/EXT-17-43188, Idaho Falls, 2017
18. T.A. Ulrich, R. Lew, R.L. Boring, K.D. Thomas, B. Rice, C.M. Poresky, Operator-in-the-Loop Study for a Computerized Operator Support System (COSS)—Cross-System and System Independent Evaluations, September 2017, INL Report No. INL/EXT-17-43390, Idaho Falls, 2017
19. A. Al Rashdan, R. Lew, L. Hanes, C. Kovesdi, R. Boring, B. Rice, T. Ulrich, The operator study on system overviews (OSSO): Design study for digital upgrades in control rooms with and without overview screens, September 2017, INL Report No. INL/EXT-17-43423, Idaho Falls, 2017

20. R. Boring, R. Lew, T. Ulrich, Advanced nuclear interface modeling environment (ANIME): A tool for developing human-computer interfaces for experimental process control systems, in *Proceedings, Part I, of HCI in Business, Government, and Organizations – Interacting with Information Systems: 4th International Conference (HCIBGO 2017), Held as Part of HCI International 2017, Vancouver, BC, Canada, July 9–14, 2017,* ed. by F.F.H. Nah, C.H. Tan (2017). https://doi.org/10.1007/978-3-319-58481-2_1

21. M.J. Peng, H. Wang, S.S. Chen, G.L. Xia, Y.K. Liu, X. Yang, A. Ayodeji, An intelligent hybrid methodology of on-line system-level fault diagnosis for nuclear power plant. Nucl. Eng. Technol. **50**(3), 396–410 (2018)

22. J. Ma, J. Jiang, Applications of fault detection and diagnosis methods in nuclear power plants: A review. Prog. Nucl. Energy **53**(3), 255–266 (2011)

23. R.L. Boring, T.A. Ulrich, J.C. Joe, R.T. Lew, Guideline for operational nuclear usability and knowledge elicitation (GONUKE). Procedia Manufacturing **3**, 1327–1334 (2015)

24. N. Zweibaum, R.O. Scarlat, P.F. Peterson, Design of a compact integral effects test facility for fluoride-salt-cooled, high-temperature reactors. Trans. Am. Nucl. Soc. **109**, 1588–1591 (2013)

Part V
Metrics

Resilient Control System Metrics

Timothy R. McJunkin and Craig Rieger

Abstract Resilience of a system, particularly critical infrastructure, is of great interest to utilities and stakeholders. Consequences of natural or man-made events (e.g., Superstorm Sandy, the sequence of storms affecting the Caribbean and coast of the United States in 2017, and the Ukrainian power grid attack) have led to emphasis and increased interest in improving resilience. However, measurement of resilience in an absolute or relative manner has been achieved in a patchwork manner. This chapter will provide the description of metric development for an electricity distribution network and begin with a definition of resilience. That definition must meet the expectation of being understandable in its linguistic form that sets a goal that would lead to systems that would stand up to large disturbances of many possible types in a manner that is quantifiable by a set of metrics. The metrics that measure a system's absolute or relative improvement is then developed for that purpose. Many notional definitions and attempts at resilience metrics formation have been coined by venerable organizations, such as the EPRI, ICS-CERT, DHS, National Academy of Sciences, and others. Common words among these sources are the ability to withstand or resist, survive, and respond expeditiously such that operations or life returns to normal as quickly as possible. To make resilience quantifiable, a proper definition of resilience and the relationship between it and reliability will facilitate development of useful resilience metrics and resilient grid architectures. The definition that is used as the basis has been developed by researchers in the community studying resilient controls over the past decade and can be found in Rieger and Rieger, Gertman, and McQueen:

T. R. McJunkin (✉)
Idaho National Laboratory, Idaho Falls, ID, USA
e-mail: timothy.mcjunkin@inl.gov

C. Rieger
Critical Infrastructure Security and Resilience, Idaho National Laboratory, Idaho Falls, ID, USA
e-mail: craig.rieger@inl.gov

© Springer Nature Switzerland AG 2019

255

C. Rieger et al. (eds.), *Industrial Control Systems Security and Resiliency*, Advances in Information Security 75, https://doi.org/10.1007/978-3-030-18214-4_12

Introduction

Resilience of a system, particularly critical infrastructure, is of great interest to utilities and stakeholders. Consequences of natural or man-made events (e.g., Superstorm Sandy, the sequence of storms affecting the Caribbean and coast of the United States in 2017, and the Ukrainian power grid attack) have led to emphasis and increased interest in improving resilience. However, measurement of resilience in an absolute or relative manner has been achieved in a patchwork manner. This chapter will provide the description of metric development for an electricity distribution network and begin with a definition of resilience. That definition must meet the expectation of being understandable in its linguistic form that sets a goal that would lead to systems that would stand up to large disturbances of many possible types in a manner that is quantifiable by a set of metrics. The metrics that measure a system's absolute or relative improvement is then developed for that purpose. Many notional definitions and attempts at resilience metrics formation have been coined by venerable organizations, such as the EPRI [1], ICS-CERT [2], DHS [3], National Academy of Sciences [4], and others. Common words among these sources are the ability to withstand or resist, survive, and respond expeditiously such that operations or life returns to normal as quickly as possible. To make resilience quantifiable, a proper definition of resilience and the relationship between it and reliability will facilitate development of useful resilience metrics and resilient grid architectures. The definition that is used as the basis has been developed by researchers in the community studying resilient controls over the past decade and can be found in Rieger [5] and Rieger, Gertman, and McQueen [6]:

> "A resilient control system is one that maintains state awareness and an accepted level of operational normalcy in response to disturbances, including threats of an unexpected and malicious nature."

The definition provides a context that a system has some minimum operational performance that has "held up" under the disturbances and been brought back to a normal state expediently. The definition further provides that the disturbances are not of one type and that they can come in combination. Time is implicit in this definition. Resilience is neither a short-term nor long-term property. It needs to consider the continuum of time frames from prior to the impact of a disturbance through the return to normalcy. This leads to a notional graphical representation of a systems response through those time frames, shown as the disturbance and impact resilience evaluation (DIRE) curve, as shown in Fig. 1. The epochs or time frames in this continuum can be described in the "Rs" of resilience shown in the figure. The *Recon*naissance phase requires the system to understand the state of the system and forecasts or potential for threats. In this time frame, the system operators may be focused on optimal performance with respect to economic efficiency or other criteria, rather than considering the response to an unexpected disturbance.

Some disturbances can be forecasted. In those cases, the operators may be considering contingency and storing up capacity to support disruptions of external support of maintaining safety and life critical resources. However, in the general case, the disturbance has no predictability. The reserve capability to respond must be

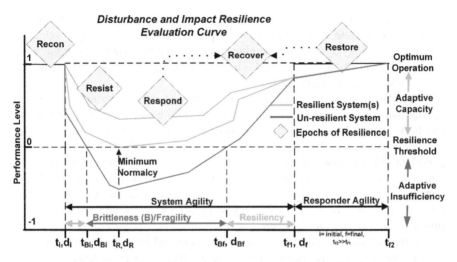

Fig. 1 The DIRE curve showing the time delineations of the 5Rs of resilience [5]

maintained at an adequate level to meet the disturbance. This reserve capability in the terminology of resilience is the *adaptive capacity* of the system. The adaptive capacity also has a temporal nature. Each asset contributing to the adaptive capacity has a latency and agility associated. Latency may be intrinsic, meaning that some amount of time is needed to turn it on. Latency may be part of the decision or the control process (e.g., we want to make sure the asset is needed before utilizing the resource). The *agility* of the system is a measure of the time the resource takes to move from the current operating point to full capability. This is like pressing down on the gas pedal of an automobile; in that, there is some ramp-up time for the full torque to be applied to the wheels.

Properties of assets that apply intrinsically to resist the disturbance comprise the *Resist* phase of the DIRE curve. This tends to be the inertial component of the system. An example of resist assets in the electric grid application include the spinning synchronous machines of generators and large motors and the available power transfer headroom in the connecting conductors and other upstream components (e.g., transformers, switch gear, relay protection). The marshland and seawall are examples for a natural system that resists storm surge. In general, the resist phase is of short duration. Assets in the resist epoch slow the disturbance, as opposed to devices that require measurement of the disturbance prior to engaging the device through a feedback loop. Such assets put to work in the next epoch have time to *Respond* before performance falls through the defined *minimum normalcy level*. The respond phase is where the rate of decrease in performance degradation is brought to a halt. A system with insufficient resist and response assets given the magnitude of the disturbance will not maintain minimum normalcy in performance and will have an *adaptive insufficiency* measured in the performance level below normalcy. The response phase would typically be longer than the resist phase.

In many cases, it may be useful to consider the notional DIRE curve time axis to be on a logarithmic scale. The following epochs of *Recover* and *Restore* are most likely successively longer time periods. Assets are brought back toward a pre-disturbance state during the recovery phase. A practical example using an automobile analogy again is the steering wheel returned to center after it has been used to avoid an obstacle in the road. This must be complete before the system will be in a state to thwart another disturbance.

Beyond recovering, a system that has been used aggressively during an event may require maintenance or even replacement in the restore phase. If a levy or pipeline has been damaged, repairs will be required to complete the cycle back to fully optimal operational performance. At the end of the restore phase, there should be a persistent effort to evaluate the system for any marginal conditions that arose. A longer-term effort to modify, refine, or add to the system should occur after larger disturbances based on lessons learned. A restore phase, assuming the calamity has been resolved, will connect restore back to the *Recon* phase with assessment and improvement as necessary.

For the metrics described in this chapter, the focus will be on the recon through the recover epochs. The application directed at electricity power systems, and more specifically modern distribution systems (MDSs), has been developed in two prior works summarized and extended here [7, 8]. The MDSs are anticipated to have different capabilities and challenges with respect to the addition of controllable generation, storage, and load. Quantitative evaluation of the response capabilities over time with respect to the magnitude and duration of the disturbance requires a like understanding of the asset and control system adaptive capacity over the time period after the disturbance. Disturbances that are of a communication and control nature are accounted for by considering increased latency or the capability that is eliminated from the attack or malfunction at the communication channel or controller level.

Modern Distribution System Resilience Metric

Methods have been developed to evaluate power systems for resilience. As the current power system evolves, introducing a greater amount of destabilizing influences from uncontrollable generation, an understanding of how resilient the resulting design is to disturbances, will be a necessity [9]. Resilience improvement in terms of distributing control loops to intelligent agents at their lowest levels with the ability to recognize precursors to faults [10] requires a measurement. Time scale of disturbance in combination with the percent of effected customers in distribution systems considering the effect on critical loads to a small set of specified disruptions was proposed as a resilience metric in Chanda et al. [11] The approach described in this chapter takes a more general and physical properties-based approach to using the definition and the notional DIRE curve mapped to the ability of the distribution system to adapt using the available assets. The previous work, focused on individual stability of active control system components for an understanding of an asset's adaptive capacity [5], is extended to the measurement of distributed assets by using a "manifold" to measure adaptive capacity across multiple dimensions including time. The metrics consider temporal flexibility of active and reactive power to measure the maximum magnitude and duration for a disturbance amplitude that can be withstood. This approach is applied to an MDS, which considers the large integration of distributed energy resources (DERs) and more complicated control mechanisms.

Stability of the power grid is defined in terms of voltage and frequency across the grid. Frequency stability requires balancing the real power (P) generation and load, combined with losses to maintain the frequency of what is a distributed electrome-chanical system buoyed by rotational inertia of the prime movers that turn the generators. Voltage stability requires the balancing of reactive power (Q) across the network. Thus, metrics must address P and Q and be extensible across the grid network. Distribution has traditionally been concerned with maintaining the connections between the distribution substation and the loads, where reliability might be enhanced from a radial network to one that contains redundancy in some paths by providing meshes with the ability to switch around branches that are out of service. MDS resilience metrics must consider the future, which is predicted to include the high penetration of DER in generation, controllable loads, storage, and other flexible assets. Control of these devices have many purposes, which include support of regulation of voltage and frequency across the distribution network, economic benefits to the owner by selling services to the grid, and the reliable utilization of interconnections (e.g., power lines, transformers, switches).

The desired outcome is a mapping of the capabilities and limitations of a distribution system to the resilient control metrics that express the "Rs" of resilience (i.e., *Recon*, *Resist*, *Respond*, *Recover*, and *Restore*, notionally in the DIRE curve represented in Fig. 1).

As described in the introduction, the magnitude and duration of a disturbance that can be withstood is dependent on the ability of a system to *Resist* and *Respond*. *Resist* describes intrinsic or immediate responding properties of the system, where *Respond* relates short latency and close proximity assets that engage automatically. To *Restore*, and to fully *Recover*, requires longer latency actions, where remote proximity of assets must be engaged at the end or near the end of the disturbance to bring the system up to optimal operation minus depleted resources. The required time for recovery is dependent on the timing and supply of the energy resource that is available to bring the system back fully to a neutral bias. The *Recover* phase includes recharging storage assets and bringing frequency and voltages to nominal values across the MDS to the extent possible. Restoring may involve repair, maintenance, or replacement of degraded assets. Prior to that repair, the assets will either be unavailable or not able to operate at full capacity. A logarithmic time scale on the DIRE curve is appropriate under most applications as *Resist* is on the order of seconds, and *Restore* may be weeks, months, or even longer.

Single-Asset Description

Assets in a distribution system need to be described in terms of power, energy, latency, and rate of change limits. When these limits can be described in a simple, yet sufficient, form, the assets can be aggregated in a manner that is tractable. This description must convey the "margin to maneuver" (M2M) that an asset has to respond to a disturbance. This description can be specified for an entire asset or the amount of an asset's capability that is set aside to respond to unanticipated changes. Like the throttle pedal example in the introduction, an example of an asset that is partially available is a generator that has been dispatched to produce 0.8 of maximum power output, leaving 0.2 of that level available in a flexible manner. The portion of the asset supporting the expected load is not counted in the adaptive capacity, since it is allocated prior to any disturbance.

Apparent Power Limits M2M is found to be in the sum of the range of control available among the generators and loads. M2M is considered for both reactive and real power. A region of the apparent power plane over which it can be applied represents the range of a component. From an analysis and design perspective, the range and any discrete steps of the choice in control are tabulated. For example, a distribution static synchronous compensator (DSTATCOM) may be adjusted continuously, whereas a capacitor bank is either on/off or has discrete steps as individual sets are switched in and out.

Energy Limits Some assets are energy limited. The battery is the simplest example to describe. A half-full battery can move either direction to supplying or absorbing power, but is limited by the storage capacity on one end and the depleted battery on the other. Buildings are another example in that they have the ability to have a temperature set point one direction or the other to accomplish a similar result in reducing load or increasing load. The building tenants likewise will not be adversely affected by small changes; however, once the building hits a steady state, there is no more power compensation available without unwanted discomfort to personnel and/or equipment. Both show example of resources that could be controlled up to a maximum power level, but only for a finite amount of time. Until the asset is recharged in a recover phase, the resource cannot be used again. The power adaptive capacity must go to zero once the energy available has been expended.

Temporal Constraints Asset capabilities can be described in two terms: latency and agility. The agility can be described as a time constant of the devices' ability to move from one output to the next or simplified as the time required to ramp up to the maximum value. The energy dynamics is implicit in the real power sourcing the energy to the grid. Once the energy store limit is depleted, the real power will be zero. The full description of the S-plane manifold specifies the maximum adaptive capacity over time.

The parameters required to calculate the limits of the asset, k, over in the apparent power plane versus are:

- $P_k m$ and $P_k M$ – limits of real power in watts (W) level of device m for minimum and M for maximum
- $Q_k m$ and $Q_k M$ – limits of reactive power in volt-amperes reactive (VAR)
- $E_k m$ and $E_k M$ – energy limits min and max in the device in Joules (J)
- λ_k – pure latency to begin to ramp $P_k(t)$

- μ_k – pure latency to ramp $Q_k(t)$
- α_k – time to ramp real power to $P_k(t)$
- β_k – time to ramp reactive power to $Q_k(t)$

For the purpose of evaluating a system versus a disturbance, which by definition we do not get to choose the direction, the P, Q, and E values used will be that of the minimum range available. For a car steering wheel example, if the wheel position is 10%, the adaptive capacity is restricted to 90%. For a generator running at 80% of 10 MW maximum generation, the adaptive capacity is 2 MW. The apparent power limit used in generating the extents of adaptive capacity over time is limited as:

$$S_k\text{max} \leq (P_{\text{max}} + Q_{\text{max}})^{\frac{1}{2}} \tag{1}$$

A resilient control system will consider that bias away from the neutral and balance other objectives against the need to be prepared to respond to an unpredicted disturbance.

Type of Single Assets

Some examples of controllable assets are shown in the complex apparent power plane of Fig. 2. Each continuously controllable asset can be described with the parameters above. Some of the expected assets in an MDS are tabulated in this subsection:

1. Continuously variable real power assets including battery to power inverter assets that do not provide reactive support, variable loads, and generation
2. Continuously variable reactive power assets, like DSTATCOMs
3. Discrete reactive power assets, such as tap changing transformers or capacitor banks
4. Discrete real power assets, like demand response of an industrial load
5. Four quadrant elements that can control real and reactive power
6. Upstream-connected distribution or transmission lines
7. Synchronous machine inertia

Fig. 2 Complex power plane with examples of the range of control of controllable assets

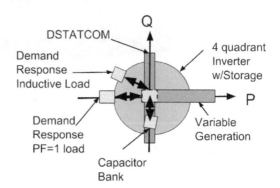

The last two classes are in some sense special cases because they are not assets that need to be controlled. With a change in the balance of generation and load, both of these will act without a man-made controller to support the frequency or voltage stability of the grid. The spinning inertia will give or take kinetic energy to the power grid as described by the swing equation, and power flow will change based on those same changes to the load generation topology. Because they are the elements that make up the resist phase of resilience, a brief development of the interaction of these two is included. For a simplified consideration, inertia can be considered as a real power asset with energy limits based on the moment of inertia and the limits on the rotational frequency and power limits based on limiting electrical and thermal characteristics of the machine that can be specified. The connecting power line asset is a non-energy constrained asset where the real and reactive power is limited by the proximity of the present load on that line with respect to the line ampacity [13] or reactance limits of the line. For further reading on determining "pinch points" on the upstream connections, refer to Parker et al. [8], where a model with expected loads in various typical loads is used to determine the tightest constraints on the electrical network.

Concise Asset Description

The asset can be described by showing the limits as a function of time from the disturbance. The bounds as time varies provide the extent an asset can be utilized in the S-plane with energy limits applied. Figure 3 shows the time-varying abilities of a

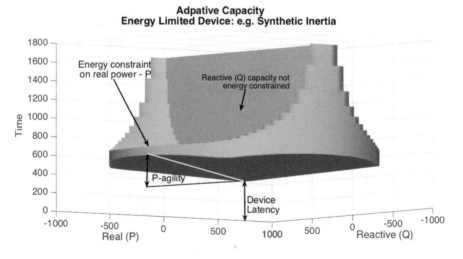

Fig. 3 A visualization of abilities of an energy-limited asset possessing both P and Q

storage asset with reactive power capabilities. With polar coordinates, the extents in the S-plane are described as a time-varying shape, $S(\theta, t)$, where θ is the angle, power factor with which the device is commanded, and t is the time from the disturbance and the application of the device; thus, the shape is stored in S in cylindrical coordinates. Since Q is not energy limited, once any P has exhausted the energy reserve, the full capabilities of Q of the device are available. A specific example of this type of asset is a power inverter that has storage and reactive elements to support both P and Q in all four quadrants of the S-plane. The shape contained in $S(\theta, t)$ is calculated for use by dividing the range of θ and t into discrete step. The surface described begins as zeros through the latency periods, the ramping in the P and Q axis at a rate based on the agility and maximum values. The magnitude at angles between the $P(t)$ and $Q(t)$ axis is limited to Smax, as given in Eq. (1).

The energy limit is determined by integrating the P component along each of the angles compared to the energy limit. At the point where the energy limit is crossed for a given θ, $S(\theta, t)$ is set to zero.

Groupings of Assets in the MDS

The assets next need to be considered as groups logically according to the topology of the connecting network. The groupings will be referred to as economic units (EcU) because their proximity put them in a common situation that is dependent on a common wire interconnecting the assets. Furthermore, the assets might together be considerable market force, whereas a single asset owner would have no negotiating leverage. A drawing is used to define the EcU in Fig. 4, which considers an aggregation of uncontrollable and controllable elements.

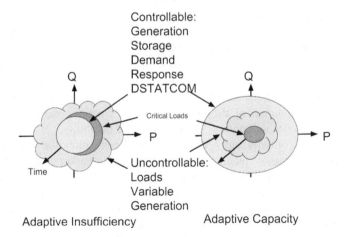

Fig. 4 Drawing illustrating the capabilities of controllable assets versus uncontrollable elements and critical loads

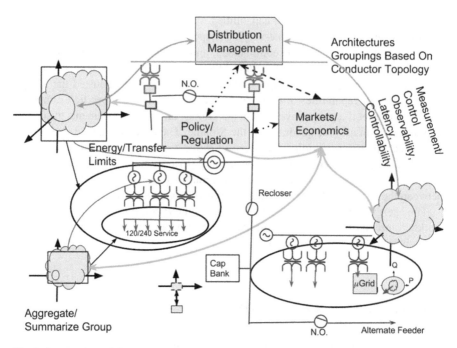

Fig. 5 Rollup of M2M in a partial primary mesh distribution system

The drawing in Fig. 4 portrays the concept of controllable versus uncontrollable aspects of aggregation. The cloud portion of the figure illustrates the variable aspects, while the solid element represents the available extents of the controllable portion without considering time in the response. The cloud portion represents loads and uncontrollable generation. These are the factors that determine the expected adaptive capacity of power and energy pipelines comprised of the conductors and other limiting components of energy transfer through the topology of the distribution network. This notional representation also shows the difference between systems that has an adaptive capacity and an adaptive insufficiency and provides a possible allocation for operational normalcy, with the delineation of a critical load. Figure 5 illustrates some logical sets of MDS assets and the aggregation of those assets including a representation of the limits, control interaction, and markets. Some elements and branches will possess greater degree of flexibility.

At the extremes of daily and seasonal loads and generation, the structure of the distribution elements is at "pinch points," as described in Parker et al. [8] To begin, the controllable assets will be account-considered. Later in the chapter, we consider the expected limits of the system at those extremes. Note the deviation of uncontrollable assets beyond the expected extremes, or at unanticipated times, will be one type of disturbance to which the adaptive capacity is applied.

The EcU is an aggregation of a portion of the MDS that for the purpose of this chapter relates to the physical topology. The EcU is supported by the upstream grid connection. Ownership or markets may group controllable aspects of the grid; however, that is left for future work that includes details of potential market interaction with the system.

Adaptive Capacity Aggregation for an EcU A raw adaptive capacity of a grouping of assets reflects the flexibility that is present in that grouping as an economic unit or an aggregation of a cut section of the power system. Adaptive capacity of a group of elements should also be considered in the context of the topological position in the distribution network. We assume the connection to the EcU is a "stiff" source (i.e., dynamics in this aggregation is not large enough to change the boundary condition voltage, and the transmission asset is additive to the controllable asset power domain). The measurement and compensation for any voltage disturbance is an operational consideration, which is neglected as metrics are used in design. Considering the limits in the S-plane, the limits on any continuously variable asset is represented as a closed contour similar to the cartoons mapped onto the S-plane previously shown in Fig. 4. The limits of the assets can be summed by expressing in polar coordinates and adding the magnitude for common elements by describing the limit as $S_{ck}(\theta, t)$ and summing the relevant assets in the total adaptive capacity in the real and reactive power is:

$$S_{AC}(\theta, t) = \sum_{k=1}^{N} Sc_k M\,(\theta, t) \tag{2}$$

over N assets in the domain, one of which is the upstream interconnect limit, $S_{cup}M$ (θ, t). Assets with discrete control (e.g., capacitor banks) are treated as a special case in the aggregation along the Q axis.

Resilient Metrics in the Context of the Grid Topology

The topology is now considered with respect to adaptive capacity and agility of the EcU and the ability for that capability to be exported outside the EcU.

Adaptive Capacity Analysis can now be made of the sufficiency of the adaptive capacity by considering the range of historical or expected noncontrollable assets in the aggregation. The shape of the cloud is the maximum magnitude of that history mapped out radially in the S-plane, S_n, forming the tangible form of the "cloud" in Fig. 4. The net adaptive capacity described in the asset aggregation manifold in Eq. (4) is:

$$S_{ACnet}(\theta, t) = S_{AC}(\theta, t) - S_{max}n(\theta), \tag{3}$$

$$S_{max}u(\theta) = \max_{t}\ S_n(\theta, t), t \tag{4}$$

defining the historical or anticipated maximum of the elements that are noncontrollable. The minimum across all angles, if it is greater than zero, then defines the margin of this portion of the system that is available before hard curtailment of load, or scheduled generation would be the recourse for mitigating the disturbance.

By excluding the upstream component from the total adaptive capacity, the exportable adaptive capacity of those assets is:

$$S_{ACex}(\theta, t) = S_{AC}(\theta, t) - S_{cup}M(\theta, t) \tag{5}$$

This neglects the loss of the export that is dependent on the resistance of the line, Ru, which would be dependent on where in the distribution system the assets' adaptive capacity is required. Remote assets would have greater losses in real power and limited voltage regulation support from the reactive elements beyond the terminal node of the EcU. The combination at the boundary, ignoring the loss, can then be added together with other branches that attach at that bus, in a similar manner as Eq. (6) at the next level of aggregation.

Agility Agility is the rate at which the adaptive capacity can be utilized. The agility in the response to the disturbance is thus the slope from the combined asset for any given EuC or aggregation of assets and EuCs. It can be expressed as the time derivative of the adaptive capacity manifold as:

$$Ag_s(\theta, t) = \frac{dS(\theta, t)}{dt}. \tag{6}$$

The derivatives provide the instantaneous agility where the extreme values of the derivatives in historical or worst-case scenarios establish the limiting levels of agility at the critical points. The mathematical expressions for agility relate to the combined rates at which assets can be applied.

Example Aggregation of Assets in a Distribution System

As described in previous sections, the adaptive capacity of the assets can easily be combined. Figure 6 shows an example of some arbitrary assets that are combined. The graphic describes the flexibility of the asset or system of assets assuming the assets are applied at the maximum rate at specified power factor angle up to the maximum apparent power level of the specific real or reactive limit is reached. Initially, a MATLAB script was written to generate the shape for all θ (i.e., power factor that can be applied to a disturbance for assets that are neutrally biased with respect to maximum adaptive capacity of real or reactive power). Subsequently, a tool for data entry and visualization was written in the Unity 3D game development platform.

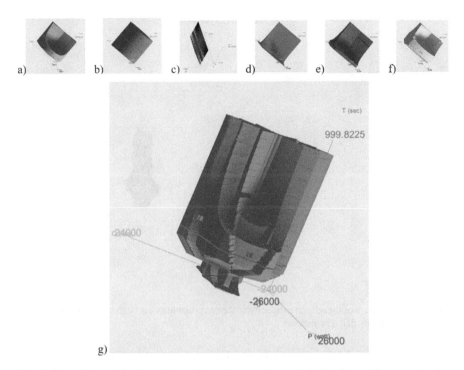

Fig. 6 Example combination of a variety of assets shown individually and in sum. Assets: (**a**) energy-limited four quadrants (e.g., battery), (**b**) non-energy-limited upstream transmission line, (**c**) DSTATCOM, (**d**) rotational inertia of a synchronous machine, (**e**) synthetic inertia, (**f**) reserve generation (e.g., spinning reserve), and (**g**) the aggregation of these assets

An example has been created in Parker et al. [8] where the limits of the connective electrical network are found to find the intrinsic resilience within distribution system. A model using load profiles for different customers can show the points at which the network is close to limits that would cause failure or trip protection. The distance away from that net trip point is a measure of the intrinsic resilience. The red line shows the extent of the usage of the connected conductor juxtaposed to the limit of the conductor in the dark blue curve shown in Fig. 7. An adaptive capacity manifold of the controllable elements is then superimposed on this extent to show the ability of those controllable assets to add to the adaptive capacity. The extended adaptive capacity is shown by the distance that the controllable assets can be engaged to move away from the edge at the inception of the disturbance. Here the intrinsic adaptive capacity "plus" the controllable asset provides the measure of EAC, which is the maximum disturbance that can be inflicted without a failure or trip. In this example, the range of the assets exceeds the limit of the intrinsic capability of the power line connecting to this EuC. As shown in detail in Fig. 8, the ability to utilize the full extent of the adaptive capacity is clipped by the intrinsic limit of the connecting power line. The clipping is not a serious problem since the limiting disturbance that would utilize this capacity would be the diameter of the intrinsic limit (i.e., much

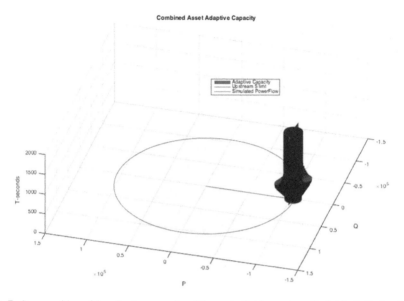

Fig. 7 Superposition of the adaptive capacity of the controllable assets on the intrinsic limits of the connected electricity delivery network

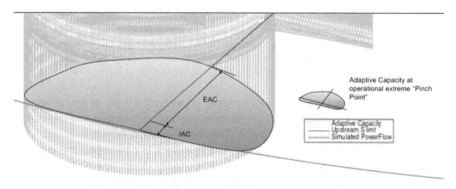

Fig. 8 Zoomed details of the clipped asset

larger than the maximum sustainable disturbance in an arbitrary direction). The graphics in Figs. 7 and 8 are a graphical representation of the analytical expression in Eq. (6).

Mapping to DIRE Curve

A direct mapping of the manifold describing the adaptive capacity to the DIRE curve is now possible. The end result is the determination of the magnitude and duration step function disturbance in an arbitrary direction in power factor. The limiting case being the nearest direction of aggregation to the intrinsic limits the resistance phase. The maximum sustainable duration for the disturbance is determined by energy limits available to continue to support the limiting magnitude, the response phase. Once the disturbance is resolved, the amount of energy available to replenish the energy reserves to levels prior to the disturbance can be assessed by determining the amount of time needed to do so in the recover phase. Finally, the restore phase will pertain to any degradation caused in the disturbance or response to assets or infrastructure that requires maintenance, repair, or replacement. The mapping of the set of the assets in the extreme of the historical or predicted worst case with respect to the limits of upstream support is shown in Fig. 9.

In the context of the power grid, the Rs of resilience are summarized:

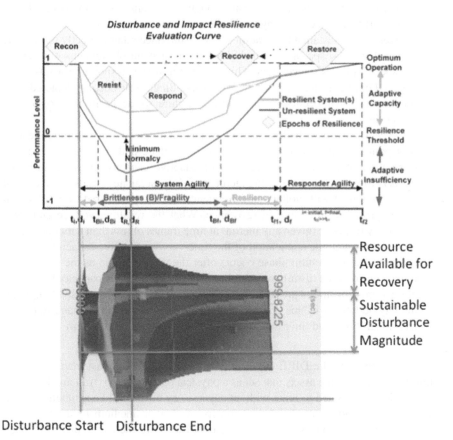

Fig. 9 Mapping of adaptive capacity manifold to the DIRE curve. The magnitude of disturbance that can be withstood is the narrowest waist of the manifold during the duration of the disturbance. The capacity remaining outside that waist and in the region right of the end of the disturbance is available for recovery. Restoration is in a time frame beyond the applicability of the power and energy resources shown in the manifold

1. *Recon:* Reconnaissance is prior to the disturbance. The adaptive capacity of the assets needs to be maintained in as complete a state as possible since the disturbance is not predictable. The better the system or operators know the state and limits of the situation, the better the response may be in the event of a large disturbance. Ideally, neutral biases of the assets are maintained by the overall control system, since choosing the direction of the disturbance is possible by definition. A neutral bias would include storage sources at a midpoint of the specified range and frequency and voltages at very near nominal specification.

2. *Resistance:* Resistance is a rate of change metric, where intrinsic properties such as inertia, fast responding assets, and acceptance tolerance to nominal voltage and frequency are considered. These determine the slope of the DIRE curve and relate to how much time slower responding assets have to react. Avoiding frequency stability "trips" to machines requires that the instantaneous support from connected spinning inertia is great enough to remain above frequency limits until short latency devices engage [12].

3. *Response:* The sustainable duration of the disturbance relates directly to the time period that the shape of the capabilities curve or manifold remains outside the cylinder of the resistance level. This allows for slower responding assets to be engaged before intrinsic and immediate responding assets become depleted. The point where that occurs would cause a failure to occur in the connecting power line.

4. *Recover:* Recovery is the longer-term evaluation of the system's ability to restore bias points in the adaptive capacity in the S-plane and the energy field. An integration of the time consideration of adaptive capacity of the disturbance impacts from $t = 0$ out to the engagement of shortest latency assets in the control space combined with voltage and frequency tolerance and credit for rotational inertia provides a measure of resistance. What the resulting voltage and local frequency response (given any inertia) at time frames of less than a minute can be analyzed through the application of the time-varying adaptive capacity.

5. *Restore:* The restoration phase occurs after the recovery epic as the system has moved on from the disturbance and has brought the adaptive capacity back to as close as possible to the pre-disturbance condition. The Restore phase, though not directly addressed in this chapter, anticipates degradation or damage requiring additional time and investment to completely return the system to optimal operation.

The mapping to the DIRE curve is illustrated in Fig. 9. Resist mapping to the intrinsic support of the assets that act as a physical process or very fast and low level control loops designed to react without supervisory decision, like synthetic inertia. The Restore period consists of the capacity of short to medium time period use. Finally, Recover is the portion that is required to bring the assets that have been depleted back to bias points where they would again be available to respond to another disturbance. Total P, Q, and E that can be applied over a given time period define the magnitude and duration of disturbance that can occur without dropping below minimum normalcy, as defined by the stakeholders.

Cyber/Communication/Control Effects on the System

Thus far, we have presented a development of resilient control metrics as applied to the operation of a modern distribution system, which implies more flexible assets and potentially complex distribution control methodology.

The method may also be used to evaluate the influence of cybersecurity and humans-in-the-loop through the application of effects in those domains in future work. For example, in cybersecurity, the potential effects of multiple types of impacts can be addressed:

- Disruption of communication channel – asset becomes uncontrollable (see Fig. 10)
- Control system compromise – asset control law set to do opposite of command
- Denial of service – increased latency (see Fig. 11)

The impact of cyber can be explored with the same manifold used to show asset power and energy capabilities. Consider the control communication channels in

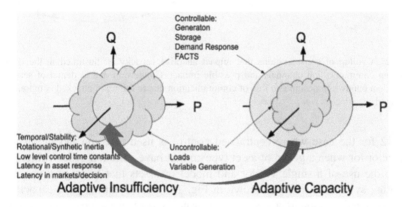

Fig. 10 The impact of a security breach of a controller within the distribution network has an impact on the adaptive capacity of that segment moving controllable assets to an uncontrollable domain or, worse, changing the sign on a feedback loop

Fig. 11 Cybersecurity may impact latency and move adaptive capacity out in time due to greater delay in activation. Denial of service or nondeterministic path for communication may cause this effect

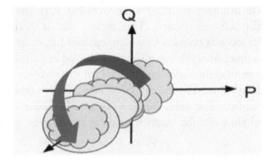

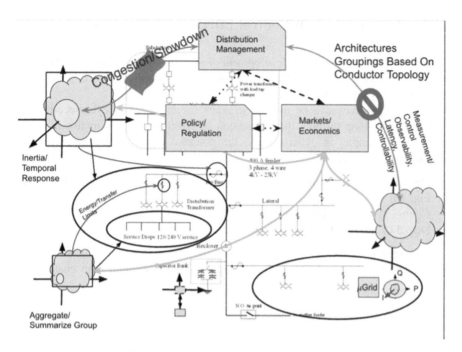

Fig. 12 A rollup of considerations that impact adaptive capacity is illustrated in the diagram depicting communication channels and possible impact. Congestion due to denial of service or insufficient bandwidth design and loss of communication due to failure or attacked component are illustrated

Fig. 12 for the case where centralized control is used to command the assets to engage or for when a group of asset types might have common cyber-vulnerability due to the use of a single vendor and model or assets that leverage a ubiquitous operating system. In the case shown in Fig. 13, a sufficient adaptive capacity has been achieved to support a system disturbance shown in the cross section on the right-hand side of Fig. 13a. For this example, we will define that enough of the diameter at the time has the minimum cross section. Now, we will assume that all of the synthetic inertia resource has been disabled through either the exploitation of the vulnerability or by cutting or jamming the communication channel to these devices. The resulting manifold is shown in Fig. 13b. The result is that the support at that time has been eliminated. This means that a disturbance of any magnitude lasting 25 seconds could not be compensated for, even though there is ample support later in time. Analysis can also be conducted in a similar example of adding latency to the control communication channel for this or other devices that could be susceptible to delays caused by a denial of service attack or other bottlenecks in the communication channel. The delay has the potential to move the asset out of the important time window that the assets were designed into the system to support.

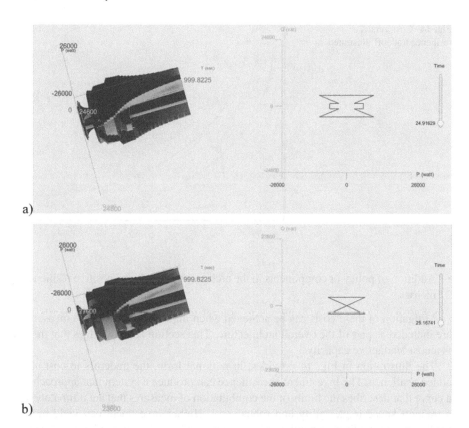

Fig. 13 A comparison of the adaptive capacity of a system of components with (**a**) expected control communication availability and (**b**) a severing of communication to short time frame support from synthetic inertia added with a super capacitor and power electronics

The Cost of Resilience

A brief section on the consideration of the cost of resilience follows. The cost of resilience includes multiple factors and not just the expense for new equipment and design. Cost is also a factor influencing DER use and the benefit to the overall architecture. That is, can it provide a correcting mechanism for behaviors that positively or negatively influence MDS resilience? As an example, the availability uncertainty for a certain type of generation or storage asset is associated with its overall benefit. In addition, the buying or selling of power from small generators requires a fairness regimen, where each asset owner is provided equal consideration. Additional costs that may also occur include:

- De-optimization of performance to maintain capability to respond. Additional sensing and control layer
- Additional information for human operator to consume

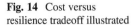

Fig. 14 Cost versus resilience tradeoff illustrated

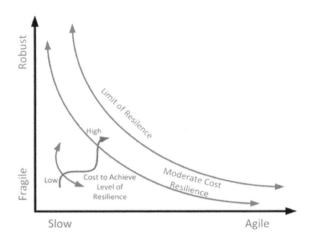

- Addition of policy or components in the architecture to achieve positive resilience metrics

Mitigation of these costs can be achieved when improved performance measures are included as part of the overall architecture. This section acknowledges that there is a cost $/adaptive capacity.

The illustration in Fig. 14 provides, in notional form, the tradeoffs in cost and added resilience. The investment in resilience can produce a system that approaches a curve that describes the limits of the combination of measures that are ultimately in a tradeoff space (e.g., agility and robustness). Higher investment can yield more resilience, but the limit of resilience is only approached with an infinite investment.

Conclusions

This chapter on resilience metrics for power systems has focused on distribution systems. The elements definition that was formed by the resilient controls community on which the metrics are rooted is provided for a general system. A concise method for accounting for the adaptive capacity that leads to resilience of the system at short and medium time frames is developed. The result of the accounting is a time versus capabilities of real/reactive power and energy that can be applied to a disturbance. The definition and metrics lead to a practical and usable description of a systems resilience measure of the magnitude and duration of disturbance that can be absorbed by the system without losing minimum performance. Additionally, the remaining resources available determine the time that it requires the system to recover to its original state after the disturbance. Those remaining resources are the power and energy available after the original disturbance has been resolved or the resource is again available, but not used during the disturbance response. The distribution of the assets across the system was described in terms of economic units that combine as a subsystem, which is of sufficient capability to have an appreciable role to play in the support of the overall system. The metrics are mapped to the DIRE curve as a tangible instance of the notional concept illustrated in the curve. Communication and control disturbances caused by failures or malicious actors are shown in a simple case study to illustrate a key concept of removing support in a particular time range. Finally, a consideration of the cost versus benefit of adding resilience is included to emphasize the fact that the investment is required to increase the resilience of the system.

Acknowledgment The authors would like to thank Jeffry Taft for his valuable conversations on distribution architectures and feedback on the development of these metrics. We also thank the US Department of Energy Grid Modernization Laboratory Consortium for funding the metric development as part of the Distribution Control Theory project.

References

1. Electric Power Research Institute (EPRI), "Grid Resiliency" webpage, n.d. https://www.epri.com/#/pages/sa/grid_resiliency?lang=en
2. A. Clark-Ginsberg, What's the difference between reliability and resilience? Stanford University, n.d. https://ics-cert.us-cert.gov/sites/default/files/ICSJWG-Archive/QNL_MAR_16/reliability%20and%20resilience%20pdf.pdf
3. Interagency Security Committee, Presidential Policy Directive 21 Implementation: An Interagency Security Committee White Paper, February 2015. https://www.dhs.gov/sites/default/files/publications/ISC-PPD-21-Implementation-White-Paper-2015-508.pdf
4. National Academies of Sciences, Engineering, and Medicine, in *Enhancing the Resilience of the Nation's Electric System*, The National Academies Press, Washington, DC, pp. 1–4, 2017. https://doi.org/10.17226/24836. https://www.nap.edu/catalog/24836/enhancing-the-resilience-of-the-nations-electricity-system
5. C.G. Rieger, Resilient control systems: Practical metrics basis for defining mission impact, in *Proceedings of the 7th International Symposium on Resilient Control Systems (ISRCS 2014)*, Denver, August 19–21, 2014, pp. 1–10, 2014. https://doi.org/10.1109/ISRCS.2014.6900108
6. C.G. Rieger, D.I. Gertman, M.A. McQueen, Resilient control systems: Next generation design research, in *Proceedings of the 2nd Conference on Human Systems Interactions (HSI 2009)*, Catania, May 21–23, 2009, pp. 632–636, 2009. https://doi.org/10.1109/HSI.2009.5091051. http://ieeexplore.ieee.org/stamp/stamp.jsp?tp=&arnumber=5091051&isnumber=5090940
7. T.R. McJunkin, C.G. Rieger, Electricity distribution system resilient control system metrics, in *2017 Resilience Week (RWS)*, Wilmington, DE, September 18–22, 2017, pp. 103–112, 2017. https://doi.org/10.1109/RWEEK.2017.8088656. http://ieeexplore.ieee.org/stamp/stamp.jsp?tp=&arnumber=8088656&isnumber= 8088637
8. W. Parker, B.K. Johnson, C.G. Rieger, T.R. McJunkin, Identifying critical resiliency of modern distribution systems with open source modeling, in *2017 Resilience Week (RWS)*, Wilmington, DE, September 18–22, 2017, pp. 113–118, 2017. https://doi.org/10.1109/RWEEK.2017.8088657. http://ieeexplore.ieee.org/stamp/stamp.jsp?tp=&arnumber=8088657&isnumber=8088637
9. V. Vittal, The impact of renewable resources on the performance and reliability of the electricity grid. *The Bridge, National Academy of Engineering* **40**(1), 351–361 (2010)

10. K. Eshghi, B.K. Johnson, C.G. Rieger, Power system protection and resilient metrics, in *2015 Resilience Week (RWS 2015)*, Philadelphia, August 18–22, 2015, pp. 1–8 (2015)
11. S. Chanda, A.K. Srivastava, M.U. Mohanpurkar, R. Hovsapian, Quantifying power distribution system resiliency using code based metric, in *2016 IEEE International Conference on Power Electronics, Drives and Energy Systems (PEDES)*, Trivandrum, Kerala, India, December 14–17, 2016, pp. 1–6, (2016)
12. J. Grainger, W.D. Stevenson, *Power system analysis*, New York (McGraw-Hill, 1994)
13. B.P. Bhattarai, J.P. Gentle, T.R. McJunkin, P.J. Hill, K.S. Myers, A.W. Abboud, R. Renwick, D. Hengst, Improvement of transmission line ampacity utilization by weather-based dynamic line rating. *IEEE Transactions on Power Delivery* **33**(4), 1853–1863 (2018)